ESTIMATING, COSTING AND VALUATION

AUTHOR

Dr. Sandeep Panchal
Lecturer, Department of Civil Engineering,
Government Polytechnic Mankeda,
Agra, Uttar Pradesh, India

REVIEWER

Dr. Arif Ali Baig
Professor, Department of Civil Engineering,
National Institute of Technology Warangal,
Warangal, Andhra Pradesh, India

KHANNA BOOK PUBLISHING CO. (P) Ltd.
4C/4344, Ansari Road, Darya Ganj,
New Delhi, India

All India Council for Technical Education
Nelson Mandela Marg, Vasant Kunj,
New Delhi, India

Dear Readers,

To prevent the piracy, this book is secured with HIGH SECURITY HOLOGRAM on the front title cover. In case you don't find the hologram on the front cover title, please write us to at contact@khannabooks.com or whatsapp us at +91-99109 09320 and avail special gift voucher for yourself.

Specimen of Hologram on front Cover title:

Moreover, there is a SPECIAL DISCOUNT COUPON for you with EVERY HOLOGRAM.

How to avail this SPECIAL DISCOUNT:

Step 1: Scratch the hologram

Step 2: Under the scratch area, your "coupon code" is available

Step 3: Logon to www.khannabooks.com

Step 4: Use your "coupon code" in the shopping cart and get your copy at a special discount

Step 5: Enjoy your reading!

ISBN: 978-93-5538-496-6

Book Code: DIP493EN

Estimating, Costing and Valuation

by Dr. Sandeep Panchal

First Edition: 2026

Originally Published by:

All India Council for Technical Education, Ministry of Education, GOI

Exclusively Published & Distributed by:

Khanna Book Publishing Co. (P) Ltd.
4C/4344, Ansari Road, Daryaganj,
New Delhi (INDIA)

Visit us at: **www.khannabooks.com**
Write us at: **contact@khannabooks.com**

To view complete list of books, Please scan the QR Code: KPH

Printed in Bharat.

प्रो. टी. जी. सीताराम
अध्यक्ष
Prof. T. G. Sitharam
Chairman

सत्यमेव जयते

आज़ादी का अमृत महोत्सव

अखिल भारतीय तकनीकी शिक्षा परिषद्
(भारत सरकार का एक सांविधिक निकाय)
(शिक्षा मंत्रालय, भारत सरकार)
नेल्सन मंडेला मार्ग, वसंत कुंज, नई दिल्ली–110070
दूरभाष : 011–26131498
ई–मेल : chairman@aicte-india.org

ALL INDIA COUNCIL FOR TECHNICAL EDUCATION
(A STATUTORY BODY OF THE GOVT. OF INDIA)
(Ministry of Education, Govt. of India)
Nelson Mandela Marg, Vasant Kunj, New Delhi-110070
Phone : 011-26131498
E-mail : chairman@aicte-india.org

FOREWORD

Engineers are the backbone of any modern society. They are the ones responsible for the marvels as well as the improved quality of life across the world. Engineers have driven humanity towards greater heights in a more evolved and unprecedented manner.

The All India Council for Technical Education (AICTE), have spared no efforts towards the strengthening of the technical education in the country. AICTE is always committed towards promoting quality Technical Education to make India a modern developed nation emphasizing on the overall welfare of mankind.

An array of initiatives has been taken by AICTE in last decade which have been accelerated now by the National Education Policy (NEP) 2020. The implementation of NEP under the visionary leadership of Hon'ble Prime Minister of India envisages the provision for education in regional languages to all, thereby ensuring that every graduate becomes competent enough and is in a position to contribute towards the national growth and development through innovation & entrepreneurship.

One of the spheres where AICTE had been relentlessly working since past couple of years is providing high quality original technical contents at Under Graduate & Diploma level prepared and translated by eminent educators in various Indian languages to its aspirants. For students pursuing 2nd year of their Engineering education, AICTE has identified 87 books, which shall be translated into 12 Indian languages - Hindi, Tamil, Gujarati, Odia, Bengali, Kannada, Urdu, Punjabi, Telugu, Marathi, Assamese & Malayalam. In addition to the English medium, books in different Indian Languages are going to support the students to understand the concepts in their respective mother tongue.

On behalf of AICTE, I express sincere gratitude to all distinguished authors, reviewers and translators from the renowned institutions of high repute for their admirable contribution in a record span of time.

AICTE is confident that these outcomes based original contents shall help aspirants to master the subject with comprehension and greater ease.

(Prof. T. G. Sitharam)

DEDICATION

This book is dedicated to my son, Shivansh, and I hope that this book will motivate him in the future as he motivates me to write.

ACKNOWLEDGMENT

I am grateful to the authorities of AICTE, particularly Prof. (Dr.) T G Sitharam, Chairman; Dr. Abhay Jere, Vice-Chairman, Prof. Rajive Kumar, Member-Secretary, Dr. Sunil Luthra, and Reena Sharma, Hindi Officer Training and Learning Bureau for their planning to publish the books on Estimating, Costing and Valuation. I sincerely acknowledge the valuable contributions of the reviewer of the book Prof (Dr) Arif Ali Baig, Professor, Department of Civil Engineering, NIT Warangal for his contribution in the form of his vast knowledge and experience on the subject.

I want to express my deepest gratitude to my **parents Mr. Chander Pal and Mrs. Sheela Devi** for their belief in me. Their unwavering support is the foundation for all my endeavors.

To my wife, Mrs. Anjali, your patience and understanding makes every achievement sweeter for me. Your continuous support and encouragement make every challenge easy for me.

To my beloved son, Shivansh Panchal, you are the light of our life. Your smile and curiosity about every small and big thing inspire me to work hard. Your laughter and imagination are always a source of motivation for me.

I would like to express my gratitude to Dr. Amit Kumar Shrivastava, Professor, Delhi Technological University, Dr. Raman Parti, Professor, NIT Hamirpur, and Dr. Chander Prakash, Associate Professor, NIT Hamirpur for their valuable guidance and expertise that shaped my writing skills time to time.

I would like to offer special thanks to Dr. Rinku Kumar, Dr. Sudhanshu Singh, Mr. Shree Krishna, Mr. Siddharth Singh, Mrs. Kiran, Mrs. Kareena, Ms. Shivani, Mr. Jatin Panchal, and Mrs. Simran to provide moral support, encouragement and helped us in every possible way.

This book is an outcome of various suggestions of AICTE members, experts, and authors who shared their opinions and thoughts to further develop engineering education in our country. Acknowledgments are due to the contributors and different workers in this field whose published books, review articles, papers, photographs, footnotes, references, and other valuable information enriched me at the time of writing the book.

Dr. Sandeep Panchal

PREFACE

The subject "Estimating, Costing and Valuation" is one of the most important subjects in civil engineering. A thorough understanding of estimating and costing is an asset for civil engineering students and practicing engineers. Accurate estimates help the stakeholders in making the right decisions on the construction project. The budget allocation by the government departments for the different projects is also based on detailed estimates. The students who have good knowledge of estimating and costing can choose their career as Estimators or Quantity Surveyors. Keeping in mind the purpose of wide coverage as well as to provide essential supplementary information, the topics recommended by AICTE are included in a very systematic and orderly manner throughout the book. Efforts have been made to explain the fundamental concepts of the subject in the simplest possible way.

During the process of preparation of the manuscript, different standard textbooks from the National and International publishers are referred. As the subject is based on the practical aspect of civil engineering, the realistic problems are discussed in detail with solved numerical examples. The book covers all types of details of estimating and costing of construction projects required for diploma-level students.

The unsolved questions included in this book are designed to make the students think about the problem. It will help in boosting the problem-solving skills of the students. The students will find it interesting to read this book. The information presented in this will enhance the curiosity among the students and will develop their interest in estimation and costing. The provisions of Indian Standard Codes, the latest rates, and recent software are discussed in detail which will help professional engineers in updating their knowledge on the subject. The book is prepared to build a strong foundation of knowledge on the subject of estimating and costing. The presented book will prepare engineering students to apply their knowledge of estimating and costing to solve the problems in planning and management of construction projects.

I sincerely hope that the book will inspire civil engineering students and professionals to learn and discuss the concepts and ideas of the subject of estimating, costing, and valuation. Further suggestions and comments are welcome for the improvement of the book for its upcoming editions. It is my pleasure to place this book in the hands of the teachers, students, and civil engineering professionals. Working on different aspects covered in the book was a big pleasure.

Dr. Sandeep Panchal

OUTCOME BASED EDUCATION

For the implementation of an outcome-based education, the first requirement is to develop an outcome- based curriculum and incorporate an outcome-based assessment in the education system. By going through outcome-based assessments evaluators will be able to evaluate whether the students have achieved the outlined standard, specific, and measurable outcomes. With the proper incorporation of outcome-based education, there will be a definite commitment to achieve a minimum standard for all learners without giving up at any level. At the end of the program running with the aid of outcome-based education, a student will be able to arrive at the following outcomes:

PO1. Engineering knowledge: Apply the knowledge of mathematics, science, engineering fundamentals, and an engineering specialization to the solution of complex engineering problems.

PO2. Problem analysis: Identify, formulate, review research literature, and analyze complex engineering problems reaching substantiated conclusions using first principles of mathematics, natural sciences, and engineering sciences.

PO3. Design / development of solutions: Design solutions for complex engineering problems and design system components or processes that meet the specified needs with appropriate consideration for the public health and safety, and the cultural, societal, and environmental considerations.

PO4. Conduct investigations of complex problems: Use research-based knowledge and research methods including design of experiments, analysis and interpretation of data, and synthesis of the information to provide valid conclusions.

PO5. Modern tool usage: Create, select, and apply appropriate techniques, resources, and modern engineering and IT tools including prediction and modeling to complex engineering activities with an understanding of the limitations.

PO6. The engineer and society: Apply reasoning informed by the contextual knowledge to assess societal, health, safety, legal and cultural issues and the consequent responsibilities relevant to the professional engineering practice.

PO7. Environment and sustainability: Understand the impact of professional engineering solutions in societal and environmental contexts, and demonstrate the knowledge of, and need for sustainable development.

COURSE OUTCOMES

After completing this course, students will be able to:

CO-1: Select modes of measurement for different items of work.

CO-2: Prepare an approximate estimate of civil engineering works.

CO-3: Prepare detailed estimates of civil engineering works.

CO-4: Use relevant software for estimating the quantities and cost of items of work.

CO-5: Justify the rate for given items of work using rate analysis techniques.

Course Outcomes	Expected Mapping with Programme Outcomes (1- Weak Correlation; 2- Medium correlation; 3- Strong Correlation)						
	PO-1	PO-2	PO-3	PO-4	PO-5	PO-6	PO-7
CO-1	3	2	2	1	1	2	-
CO-2	3	3	3	2	2	3	1
CO-3	3	3	3	2	2	3	2
CO-4	3	2	2	1	3	2	2
CO-5	3	3	3	2	2	3	-

ABBREVIATIONS AND SYMBOLS

List of Abbreviations

General Terms			
Abbreviations	**Full form**	**Abbreviations**	**Full form**
PWD	Public Works Department	GL	Ground Level
IS	Indian Standard	Pa	Plinth Area
BIS	Bureau of Indian Standards	CC	Cement Concrete
RCC	Reinforced Cement Concrete	LC	Lime Concrete
DPC	Damp Proof Course	CI	Corrugated Iron
AC	Asbestos Concrete	GI	Galvanized Iron
OPC	Ordinary Portland Cement	RB	Reinforced Brick
DPC	Damp Proof Course	WC	Water Closet
CBRI	Central Building Research Institute	POP	Plaster of Paris
		BBS	Bar bending schedule
C_l	Cost of Land Acquisition	FSL	Full Supply Level

List of Units and Symbols

Symbols/ Units	Description	Symbols/ Units	Description
B	Breadth	%	Percentage
H	Height	gm	Gram
W	Width	*cc*	Cubic centimeters
Φ	Diameter of bar	*sq m*	Square metre
R	Radius of arch	*cu m*	Cubic metre
mm	Millimeter	₹	Indian rupee
Kg	Kilogram	T	Tera
cu m	Cubic meter	M	Mega
MPa	Mega Pascal	G	Giga
m^2	Square metre	d	Deci
m	Metre	J	Joule
°	degree	*mm*	Millimeter
Kg/m^3	Density	*nos*	Numbers
mm	Millimeter		

LIST OF FIGURES

LIST OF TABLES

GUIDELINES FOR TEACHERS

To implement Outcome Based Education (OBE) knowledge level and skill set of the students should be enhanced. Teachers should take a major responsibility for the proper implementation of OBE. Some of the responsibilities (not limited to) for the teachers in OBE system may be as follows:

- Within reasonable constraint, they should manoeuvre time to the best advantage of all students.
- They should assess the students only upon certain defined criterion without considering any other potential ineligibility to discriminate them.
- They should try to grow the learning abilities of the students to a certain level before they leave the institute.
- They should try to ensure that all the students are equipped with the quality knowledge as well as competence after they finish their education.
- They should always encourage the students to develop their ultimate performance capabilities.
- They should facilitate and encourage group work and team work to consolidate newer approach.
- They should follow Blooms taxonomy in every part of the assessment.

Bloom's Taxonomy

Level	Teacher should Check	Student should be able to	Possible Mode of Assessment
Create	Students ability to create	Design or Create	Mini project
Evaluate	Students ability to justify	Argue or Defend	Assignment
Analyse	Students ability to distinguish	Differentiate or Distinguish	Project/Lab Methodology
Apply	Students ability to use information	Operate or Demonstrate	Technical Presentation/ Demonstration
Understand	Students ability to explain the ideas	Explain or Classify	Presentation/Seminar
Remember	Students ability to recall (or remember)	Define or Recall	Quiz

GUIDELINES FOR STUDENTS

Students should take equal responsibility for implementing the OBE. Some of the responsibilities (not limited to) for the students in OBE system are as follows:

- Students should be well aware of each UO before the start of a unit in every course.
- Students should be well aware of each CO before the start of the course.
- Students should be well aware of each PO before the start of the program.
- Students should think critically and reasonably with proper reflection and action.
- Learning of the students should be connected and integrated with practical and real-life consequences.
- Students should be well aware of their competency at every level of OBE.

CONTENTS

1 Fundamentals of Estimation and Costing

UNIT SPECIFICS

In this unit, the following aspects will be discussed:

- Meaning and purpose of estimation and costing
- Various types of approvals
- *Types of estimates and their uses*
- The standardized formats of estimation sheets
- Modes of measurement and deductions in different works
- Specifications of work items

To enhance the student's practical understanding, the provisions in Indian Standard Codes are discussed. Some of the important information is shared in tabular form for a better understanding of the concepts.

This unit consists of exercises with long and short-term questions, multiple-choice questions, a list of references, and suggested reading to make the students learn through practice. Some extra information that may interest the students is provided in the form of QR codes. The students can scan these QR codes to reach the source of further knowledge on the different topics.

RATIONALE

This unit on the fundamentals of estimating, costing, and valuation gives the introduction of the different concepts of estimating and costing. The approvals and sanctions required for the execution of a project are discussed. The different types of estimates are discussed in detail and their suitability in different circumstances is analysed. The checklist of items and standard formats of estimation and costing are provided in tabular format. A detailed discussion on the mode of measurement and deductions during the estimation is provided. The content presented in this unit helps in building a foundation of knowledge on the subject and develops the interest of the students further.

Estimating and costing is an important subject in civil engineering. It helps the students understand and calculate the probable cost of the construction projects. The concepts of approvals, types of estimating formats, and deduction in the items of work are essential to understand for a civil engineering student. The students will learn about the provision of the codes and their use in estimating the probable cost. The knowledge gained in this unit can be used by the students in their project work, and fieldwork. The students who gain expertise in the subject of estimating, costing, and valuation may choose it as their career too.

PRE-REQUISITES

No prerequisites are required for studying this unit.

UNIT OUTCOMES

The list of outcomes of this unit is as follows:

U1-O1: Understand the different types of approvals required for the construction project

U1-O2: Explain the different types of estimates

U1-O3: Prepare the different formats of estimation

U1-O4: Analyse the deductions in different items of work

Unit-1 Outcomes	EXPECTED MAPPING WITH COURSE OUTCOMES (1- Weak Correlation; 2- Medium correlation; 3- Strong Correlation)				
	CO-1	CO-2	CO-3	CO-4	CO-5
U1-O1	2	-	-	-	-
U1-O2	3	2	-	-	-
U1-O3	3	1	-	-	-
U1-O4	3	2	-	-	-

1.1 INTRODUCTION TO ESTIMATING AND COSTING

The engineers and construction planners are required to calculate the probable cost of construction of any proposed structures. It is the duty of civil engineers and quantity surveyors to give an accurate estimate of the possible cost of construction so that it can be compared with the available funds. If the probable cost is less than the available funds, then the engineers can go on to further stages. However, if the estimated cost is more than the available funds, the engineers need to reduce the probable cost.

It is the duty of the quantity engineer to forecast the probable cost accurately. If the estimate of the probable cost is less than the actual cost of construction, the funds remain idle which is not good for large-scale industries. If the calculated probable cost is more than the actual cost of construction, the owner may face the financial crisis that may lead to the alteration of design or suspension of the work. However, if the estimate is prepared using all the possible circumstances, the probable cost and actual cost will have a very small difference. The process of calculating the quantities of the items and the probable cost involved is known as estimation.

1.2 PURPOSE OF ESTIMATION

As discussed in the previous section, two types of costs can be calculated in the construction of infrastructure. The approximate cost of construction can be calculated from the drawing and plans however, the final or actual cost of construction can be calculated only after the construction. The process of estimation and costing plays an important role in determining the approximate cost before the beginning of the construction and helps the owner understand the financial liabilities. The main purposes of the estimation and costing are discussed as follows:

(1) The approximate cost gives an idea about the feasibility of the proposed project. The probable cost of construction helps the owner to understand if one can arrange funds for such a project or not.

(2) The estimation of probable cost helps the owner in decision-making. In the case of government agencies, probable cost helps in obtaining administrative approval for the execution of the project.

(3) An estimate is required to select the construction materials and equipment. The cost of construction varies with the type of material and construction technology. So, the estimate can help in choosing the construction material and machinery for the completion of the project within available funds.

(4) The estimate helps in deciding the requirement of labour for construction. After analyzing the detailed probable cost, the engineer can decide the quantity of labor of different types.

(5) The detailed cost of construction is used to decide the timeline of the project. It also helps in controlling the cost at different stages of construction.

(6) The detailed estimates are required for inviting and finalization of the tenders. The calculation of different items of work and their cost helps in choosing the best contractor for the work.

(7) The decision-making for any commercial project depends on the cost-benefit ratio. The estimates are used in the calculation of the cost-benefit ratio.

(8) The estimate also helps for insurance purposes. It is used for the valuation of the property too. An idea of the probable cost of construction also helps in deciding the rent.

1.3 APPROVALS AND SANCTIONS

The construction work can't be started directly. There is a need to obtain some formal acceptance from the concerned departments. The major types of approvals and sanctions are discussed as follows:

(1) **Administrative Approval:** Administrative approval is required to begin the technical planning of the work. It is mandatory to get administrative approval from the competitive authority. Administrative approval may be defined as the 'formal acceptance of the proposal' by the competitive authority. At the initial stage, rough estimates and preliminary plans are required for administrative approval. In most of the cases, the administrative approval is not required for the repair works.

The detailed estimates and drawings are prepared as the administrative approval is obtained. Scan the QR code to have a look at the administrative approval of an actual project of the Government of Assam, India.

(2) **Expenditure Sanction:** The expenditure sanction may be defined as an agreement by the administration/ competitive authority on the proposed cost. Any expenditure before this sanction is not allowed for the proposed construction. The ministries and government departments possess the power for expenditure sanction.

(3) **Technical Sanction:** As the administrative approval is given by the competitive authority, the process of making detailed estimates, designs, and detailed plans is started. These detailed proposals are evaluated technically by experts in civil engineering departments like the Public Work Department (PWD) for their technical feasibility and accuracy. The formal acceptance of the detailed proposal including designs, detailed estimates, detailed plans, etc. by technical departments is known as 'technical sanction'.

(4) **Budget Provision:** After technical sanction, a report including documents like detailed designs, the details of the raw material, designs, the steps to achieve the goals, and the detailed costs are prepared. The material required and estimated cost at the different stages of the project are mentioned in the report. The budget provision helps in controlling the cost of the construction at different stages.

1.4 CLASSIFICATION OF ESTIMATES

At this stage, the students would have understood the meaning and importance of the estimate. The different types of estimates are used at the different stages of the construction. Based on the details and accuracy, estimates are classified into the following two types:

(1) Approximate Estimate

(2) Detailed Estimate

1.4.1 Approximate Estimate

Approximate estimates are the estimates that give the rough or approximate cost of the construction. The calculations in the approximate estimate are not performed in very detail. Detailed designs and plans are not required to prepare these types of estimates. Such types of estimates are prepared for making initial decisions on the project. The preliminary survey data and soil reports with a line diagram of the project can be used for the preparation of approximate estimates. The approximate estimates are required for the administrative approval of the project. The probable cost calculated in the approximate estimate should not differ from the actual cost as it helps in making the primary decision on the project. The approximate estimate is used for the following purposes:

(1) The approximate estimate helps in deciding the feasibility of the project. It helps in understanding if the project should be undertaken or not. If the approximate cost is too high as compared to the available budget, the project should be dropped.

(2) The preparation of a detailed estimate requires a lot of time and money. So, it is a general practice to prepare an approximate estimate to save time and money.

(3) The administrative approval is given based on the approximate estimate. It helps the competitive authority to make decisions on the primary approval of the project.

(4) It helps in tax calculation and the primary valuation of the property. It can also be used to decide the rent or cost of booking flats in the proposed buildings.

1.4.2 Detailed Estimates

The detailed estimate provides an accurate estimate of the probable cost of construction in a detailed manner. The cost is calculated based on the quantity of items to be used in the construction. The different items of work like plastering, paints, bricks, etc. are predicted, and their cost per unit item including the profit is also estimated. The cost is calculated by multiplying the quantity of items of work by their cost. The detailed estimate is prepared in the following two steps:

(1) Calculation of detailed measurements and computation of quantities

(2) Abstract estimate

In the first step, the details of the different items of work in the form of their dimensions and the number of items required are used to calculate the different quantities. In the abstract estimate, the cost is calculated by the quantity of the item and the rate per unit item. The approximate estimates and detailed estimates are discussed in detail in the upcoming chapters.

1.5 TYPES AND USES OF DETAILED ESTIMATES

Different types of detailed estimates are used for different purposes. The different types of estimates based on their purpose are enlisted as follows:

(1) Revised Estimate

(2) Supplementary Estimate`

(3) Revised and Supplementary Estimate

(4) Repair and Maintenance Estimate

(5) Renovation Estimate

1.5.1 Revised Estimate

A revised estimate is provided when there is a deviation in the quantity and rate of the item as compared to the estimate provided by the estimator. A comparative statement in the abstract format showing the variation of the quantity and rates is provided with the revised estimate. The revised estimate is provided when the sanctioned amount is changed due to the change of the rate of material or quantity of the material. In this type of estimate, the plans and drawings remain unchanged.

1.5.2 Supplementary Estimate

When the work is in progress, the design may be changed at some stage for the development of the project. In such cases, some independent work that is different from the sanctioned work is required. In this situation, a supplementary estimate is provided. The procedure of preparing the supplementary estimate is the same as that of the detailed estimate and the comparative statement is not required in this case. However, the reason and necessity of the deviation of the structural design is explained in the report while providing the supplementary estimate.

1.5.3 Revised and Supplementary Estimate

As discussed in the previous sub-sections, the revised estimate is prepared when the rate of items or quantity is changed and the supplementary estimates are provided when deviation in the structural design arises. Sometimes, both of these situations arise together. In this case, a revised estimate combined with the supplementary estimate is prepared. The amount of the supplementary estimate is added to the revised estimate which is shown separately.

1.5.4 Repair and Maintenance Estimate

When the work is completed, it is needed to be maintained and repaired periodically. So, an annual repair and maintenance estimate is prepared. The cost for some of the items of the works that require periodic maintenance like whitewashing, painting of doors, etc. can be predicted in the estimate. However, works like patchwork, change of glass panes, or repair of the plastering work and wall, the estimation is not possible in advance. In this case, a lump sum amount for such repair and maintenance work is added. The cost incurred for such kind of maintenance and repair of a similar project is used to get an idea of such costs.

1.5.5 Renovation Estimate

The process of improving the damaged or broken structure is known as renovation. There may be partial renovation of the structures or complete renovation in which the whole structure is reconstructed. So, renovation estimates are required for the improvement of the damaged structures. Repair and maintenance are performed at a relatively smaller scale; however, the renovation may include reconstruction of the damaged structures too. The renovation estimate is a kind of detailed budget and is prepared for the structures after using them for some years.

1.6 ROLES AND RESPONSIBILITY OF ESTIMATOR

The estimator plays an important role in the construction projects. The estimation engineer's work starts before the beginning of the project and it goes on even after the completion of the project. The different roles and responsibilities of the estimator at the different stages of the construction project are explained as follows:

(1) **Role and Responsibility at the Preconstruction Stage:** Before the beginning of the project, the estimator performs the feasibility studies to estimate the primary cost of construction. The estimation engineer analyses the conditions of the site, availability of the material, availability of the labour, and other political or social conditions that can impact the project in the future. He prepares the budget and initial cost by working in coordination with the engineer and architect.

(2) **Role and Responsibility at the Designing Stage:** The estimation engineer predicts the quantity of the material required at the different stages. The estimate of the quantity of the material is calculated from the drawing and plans provided by the engineer and architect. After the determination of the quantities of the material, the cost of the material is also calculated by the estimation engineer.

(3) **Role and Responsibility at the Bidding Stage:** The estimation engineer prepares the tender documents and bid documents. He also negotiates with the vendors.

(4) **Role and Responsibility at the Construction Stage:** The estimation engineer monitors the actual cost during the construction and compares it with the estimated cost. He also analyzes the deviation of the actual cost from the estimated cost and its impact on the project finance. He suggests solutions for this variation in the cost. Sometimes, during construction, there may be some changes in the design due to some unavoidable circumstances. The estimation engineer must update the budget according to the changes in the construction. He also works with the project manager to minimize the cost of construction.

(5) **Role and Responsibility at the Post-construction Stage:** After the completion of the project, the estimation engineer compares the total actual cost of construction and the estimated cost. He identifies the possibilities for improvements in future projects. He also completes the documentation including final costs, invoices, etc.

1.7 CHECKLIST OF ITEMS

A checklist of the items helps the estimation engineer to prepare an accurate estimate. It ensures that all the items are included in the estimate and no critical construction component is left. If some component is missed during the estimate, it will lead to an inaccurate estimate. A checklist of items gives transparency in the estimation as the stakeholders and clients can review the checklist to understand the preparation of the estimate. The checklist of items for the load- bearing structures is given in Table 1.1:

Table 1.1: Checklist of items in load-bearing and framed structures

Load-bearing Structures	*Framed Structures*
(1) Clearance of the site (2) Excavation (3) Plain cement concrete bed in foundation	(1) Clearance of the site (2) Excavation (3) Plain cement concrete bed in foundation
(4) Foundation masonry (5) Damp proof course (DPC) with cement mortar ratio of 1:3 (6) Brick masonry in cement mortar of ratio 1:6 (7) Filling of plinth (8) Flooring (9) Doors and windows (10) Plastering	(4) Plinth beam and columns (5) Floor beams and slabs (6) Filling of the plinth (7) Masonry in superstructure (8) Staircases (9) Flooring (10) Doors and windows (11) Plastering
(11) Painting and finishing (12) Plumbing and electrical works	(12) Painting and finishing (13) Plumbing and electrical works (14) Insulation materials

1.8 STANDARD FORMATS

The estimate is prepared in the form of some standard formats. These formats help the estimation engineer present calculations in a detailed and brief manner. The estimate is presented in the following types of formats:

(1) Measurement sheet

(2) Abstract sheet

(3) Face sheet

1.8.1 Measurement Sheet

The measurements of the items and their quantity are presented in the measurement sheet. The construction work is divided into different items of the work like earthwork, concreting, brick masonry work, etc. These items of the work are further classified into the sub-headings. The total quantity of the item of the work is calculated using the dimensions of the item. The standard format for the measurement sheet is shown in Table 1.2.

Table 1.2: Standard format of measurement sheet

Item No.	Description or Particulars of work	No.	Length	Breadth	Height or Depth	Content Quantity	Total Quantity	Remarks
(1)	(2)	(3)	(4)	(5)	(6)	(7)	(8)	(9)

The content quantity is calculated by multiplying the dimensions (Column 4 to 6) and the total quantity is calculated by multiplying the content quantity (Column 8) by number (Column 3).

The total quantity calculated in the measurement sheet is used for the calculation of the cost in the abstract sheet.

1.8.2 Abstract Sheet

The cost of the construction is calculated in the abstract sheet. The abstract sheet consists of the quantity, cost per unit item, and total cost. The total cost of the project is calculated by adding the cost of the total items. The cost of contingencies can be taken 3% to 5% of the total cost. The contingencies include the expenditure on uncertain items, petty works, or the cost incurred on any unpredictable condition during the project. There may be some employees like guards or accountants for the project. The cost incurred on such employees comes in the category of work charge establishment. The work charge establishment cost is taken 1.5 % to 2% of the total cost. The contingencies and work charge establishment cost are added to the total cost to calculate the final cost of the project. Table 1.3 shows a standard format of the abstract sheet.

Table 1.3: Standard format of abstract sheet

Item No.	Description or particulars of the work	Quantity	Unit	Rate	Amount	Remarks
(1)	(2)	(3)	(4)	(5)	(6)	(7)

1.8.3 Face Sheet

The face sheet briefly gives the total cost of the project. It acts as the front cover of the detailed and abstract estimate. It consists of the estimated cost, contingency charges, work-charge establishment charges, and any other charges that are not considered in the detailed estimate but can be taken as a percentage of the total cost. A standard format of the face sheet is given in Table 1.4.

Table 1.4: Standard format of face sheet

S. No.	Particulars	Amount
1	Estimated cost	
2	Water supply and sanitary charges @.........%	
3	Electrification charges @......%	
4	Contingencies @..........%	
5	Work-charge establishment @..........%	
	Total amount in Figures	
	Total amount in Words	

1.9 PRINCIPLES OF MEASUREMENT

The measurement of the items is a mandatory and important part of the estimation. The whole work is divided into the items of the work and measured based on the dimension. The major principles of the measurement for preparing the estimate are enlisted as follows:

(1) Thick works or mass voluminous works are measured in volume. The units may be cubic meters. Masonry work is an example of the thick work.

(2) The thickness of the thin or shallow works or the surface works is measured in the area or square meters. The example of such thin or shallow works is wall paints or plaster.

(3) The long work having negligible breadth and depth is measured in the running meters.

(4) The other works like piecework are considered in numbers.

(5) The arrangement for the soaking of the bricks, concrete mixing, curing of concrete, etc. are considered in the rate of work, and no separate payment is done for such works.

(6) The rate of any item of work is considered the same for all the floors unless it is mentioned in the description of the work.

1.10 MODES OF MEASUREMENT AND ACCURACY (AS PER IS 1200)

Measuring the items is a mandatory and important part of the estimation. Indian Standard Code IS 1200 provides information about the different items of work, the unit of measurement, and the mode of measurement. The different items of the work, unit of measurement, unit of payment, and mode of the work are discussed as follows:

1.10.1 Earthwork

Earthwork includes cutting and filling works during the construction. It is the process of movement of the earthwork by excavation and filling. The earthwork is measured in cu m and the unit of payment is % cu m. It means that the payment for the earthwork is done per 100 cu m of earthwork volume. The earthwork is divided into three parts as discussed below:

(1) **Excavation:** The measurement of the excavated earth is done according to the drawings. If the depth of the excavation is not more than 30 cm, the excavation is measured in sq. m. In this case, average depth is considered. The rate of the excavation varies with the soil and rocky strata.

(2) **Filling:** The filling is done with sand or loose earth. The deductions are made for considering the consolidation of the filling material. A deduction of 10% is considered in the measured volume for filling for the consolidated fill. However, if the consolidation is done with heavy machines, the deduction is kept at only 5%. For confined places like floors, there is no provision for deduction in the filling of earthwork.

(3) **Charges for shoring:** Shoring is the process of supporting the excavated portion with the help of formwork. The charges of labor and measurement of the earth-supporting items like plank, struts, timber work, etc. are considered in the estimate. The provisions for measurement of earthwork are given in IS 1200 (Part I): 1992. The students can download part I of this code through the given QR code.

1.10.2 Brickwork

The size of a modular brick is 19cm×9cm×9cm, and the nominal size of the modular brick is 20cm×10cm×10cm. The brickwork is done in the lime mortar or cement mortar. The major points to be considered in the brickwork are discussed as follows:

(1) The brickwork of one brick thickness refers to the situation when the length of the brick is parallel to the thickness of the wall. The brickwork can be performed in one brick-thick wall or more than one brick-thick wall. Lime mortar or cement mortar is used in the brickwork. In the brickwork, the area of the openings and lintels is deducted to obtain the net measurement. However, there is no need for the deduction for the small curves and chamfers. The measurement of the brickwork on the different

floors is performed separately. The deductions for the openings up to 0.1 sq. m are not required. There is no provision for extra payment for the construction of the small curves or chamfers, however, for cutting work extra payment can be made.

(2) The brickwork in the arches is measured separately and extra rates are given for the arches. Hire and labor charges for the centering and shuttering are not calculated separately up to a 6m span of arches and are included in the rate of the item.

(3) The measurement of the reinforced brickwork is done separately as compared to the general brickwork. The reinforcement is measured separately however, the methods for the deductions of the openings remain the same.

(4) The measurement and payment are done in per cm of depth for cutting holes through the existing brickwork.

1.10.3 Concrete Work

The concrete work in the different parts of the structures is measured separately. The measurement of the concrete work is discussed as follows:

(1) **Concrete in foundation:** The lime or cement concrete in the foundation work is measured on the finished work. The unit of measurement and the unit of payment is cu m. The measurement is taken to the nearest cm of the finished work. The gradation, mix design, and technique used for the mixing of the concrete components are described. The testing required on the materials or the finished work is also stated. The reinforced concrete and plain concrete are measured separately.

(2) **Reinforced cement concrete (RCC):** The measurement for the slabs, partitions, etc. are taken to the nearest 0.5 cm while the other components are measured near 1 cm. In reinforced cement concrete, the deductions are not made for the reinforcement. The deductions are not made for pipes with volumes up to 25 sq. cm and openings up to 0.1 sq. m. The unit of measurement and payment are cu m for RCC.

(3) **Damp-proof course (DPC):** The damp-proof course is measured as per the details of the drawing sheet. The DPC is measured in sq. m and the unit of the payment is also sq. m. The details of the shuttering, finishing, curing, etc. are given in the description.

(4) **Expansion joints:** The expansion joints in the roof, walls, etc. are measured in running meters. The depth and width of the joint is also stated.

(5) **Concrete blocks:** If the concrete blocks exceed 10 cm on the bed, the measurement is taken in cu m otherwise the measurements are taken in the sq. m. The deductions are not made for the hollow portion of the hollow blocks.

1.10.4 Stone Work

The stonework in the construction of the wall or on the wall facing is measured separately. The measurement in cu m is measured to the nearest 0.01 cu m, the measurement of the area is measured to the nearest 0.01 sq. m, and the linear measurement is measured to the nearest 0.01 m. The measurement of the stonework is discussed as follows:

(1) In the case of random rubble masonry, coursed rubble masonry, and ashlar masonry, the thickness of the wall is measured nearest to 1 cm for the fraction 0.5 cm or above. However, the fractions below 0.5 cm are ignored. The unit of the measurement and the unit for payment is cu m.

(2) If the stone is used in the facing of the wall, the average width of the bed and the thickness of the joint is given. If the stone for the construction of the wall and the facing are the same, extra cost for the dressing of the stone can be considered. The unit for the measurement and the unit for payment for the wall facing is sq. m.

(3) Cutting and dressing of the stones is divided according to the sizes of the stones. The labour for dressing the stone is measured separately. The unit for the payment and unit for the measurement is cu m.

(4) The provisions of the deductions for openings are the same as the brickwork. However, the stonework of different types of stones is measured separately.

1.10.5 Roof Work

The measurement of the roofing work is based on the type of roof and the material used for the roofing. The measurement of the roofing is discussed as follows:

(1) Terraced roof: The unit of measurement and the unit of payment is sq. m. The size and quality of the bricks, tiles, or stone slabs are mentioned. The technique for laying, the mix of mortar, the number of layers, etc. are also stated. The supporting construction members like beams or rafters are measured separately. If lime is used for terracing of the roof, the consolidated thickness is mentioned. For special finishes, the provision of extra payment is given. IS 1200 (Part IX): 1973 (Reaffirmed in 2007) gives the details about the roof covering. The students can download this part of code by scanning the QR code.

(2) Reinforced slab roofing: The unit of measurement and unit of payment are cu m. The steel reinforcement and shuttering are measured separately. The measurement is taken as per the drawing.

(3) **Asbestos corrugated roofing:** The measurements are taken as per drawings. The unit of measurement and unit of payment is sq. m. The sheeting is measured flat and the method of fixing is mentioned. The type of the sheet and thickness of the sheet are mentioned.

(4) **Waterproofing of roof:** The quantity of bitumen required per sq. m is mentioned. The unit of measurement and unit of payment is sq. m.

(5) **Ceiling:** The material, the thickness of the ceiling, and the technique of fixing is mentioned for the ceiling. If the opening size is below 0.4 sq. m, no deduction is required.

1.10.6 Floor Work

The measurement of the floorwork is performed in the sq. m and the unit of payment is also sq. m. The measurement of the flooring is discussed as follows:

(1) In the ground floor, the lime base and the floor finishes are taken as single items. These are measured in sq. m. However, in the upper floors, the supporting and finishes are measured separately.

(2) The thickness of the lime or cement floor is mentioned. If the thickness is more than 20 cm, it is considered under foundation works.

(3) The rounding angles in the case of artificial stones are measured in running meters. The thickness of mosaic or terrazzo flooring is measured after polishing.

1.10.7 Plastering

The measurement and payment of the plastering work is done in sq. m. The measurement of the plastering is discussed as follows:

(1) The thickness of the plastering layer is taken 12 mm, normally.

(2) The wall plastering measurement is done for the interior and exterior faces. The deduction rules for plastering are discussed in the upcoming section in detail.

(3) For the ceiling, the measurement of the ceiling area is taken for the measurement of the plastering area. However, if the ceiling with the projected beam is provided, the ceiling is measured over the beam, and the sides of the beams are measured and added to the ceiling area.

(4) The plastering operations on the ceiling are difficult to perform and the mortar mix is richer as compared to the wall plastering. However, the thickness of the ceiling plastering is also 12 mm but it is measured separately.

1.10.8 Doors and Windows

The doors and windows are divided into two parts, the frame and the leaves or shutters. The measurements of the frames and the leaves are taken separately. The measurements of the doors and windows are discussed as follows:

(1) **Frame:** The unit of measurement and unit of payment is cu. m. The length of the frame is obtained by adding the length of all the members. Length is measured to the nearest 2 cm while the width and breadth are measured to the nearest 2 mm. If a sill is provided, the length of the sill is also measured. The window frame may have more than two vertical members. The length of all the members is added in the case of the window frame. The projections, sills, etc. are considered in the measurement.

(2) **Leaves or shutters:** The unit of measurement and unit of payment is sq. m and the thickness of the leaves is mentioned. For the measurement of the breadth and the height of the leaves, the rebates of the frame are taken into consideration. The rebate may be 12 mm to 20 mm in the frame. If the sill is not provided, a clearance of 6 mm may be allowed. In the case of timber doors and windows, the quality of timber, finishing, and fittings are mentioned. Fittings are measured separately in numbers. These can be taken as a separate item showing the lump sum cost.

1.10.9 Wood Works

The type of material, finishes, and fittings required are mentioned in the specification for the woodwork. The doors and windows are already discussed and the norms remain the same for the wooden doors and windows. The type of material and finishing should be mentioned. The waste material from the wooden work is not taken into consideration. There is no provision for giving allowance for nails, screws, pins, etc. However, the hinges, bolts, etc. are considered separately. The trusses, wooden beams, posts, etc. are measured and paid in cu m, however, shutters of doors, windows, wooden trellis work, etc. are measured in sq. m. Handrails, wooden piles, etc. are measured in running meters.

1.10.10 Steel Works

The steel works and the ironwork are measured and paid by weight generally. Plain rolled sections, steel bars, etc. are measured in Kg or quintals. The steel table can be used for obtaining the weight/running m for the different sections. If a steel table is not available, the weight is calculated by considering the mild steel density of 7850 Kg/cu m. For RCC works, the weight of the reinforcement is considered and the weight of the binding wires is not considered. However, the measurement of the binding wire is included in the item.

The nuts, bolts, rivets, etc. are calculated in numbers, and their weight can be calculated using a steel table. In some cases, the nuts, bolts, and rivets are measured as the percentage of the whole steelwork. For doors and windows, rules remain the same as discussed in section 1.10.8.

1.10.11 Whitewashing and Distempering

The measurement and payment for items of whitewashing and distempering are done in sq. The area of the whitewashing and distempering is kept the same as that of plastering and no extra calculation is done. IS 1200 (Part XIII): 1994 gives the details about whitewashing and painting.

This part of the code can be downloaded by scanning the given QR code. The inside whitewashing is taken as equal to the inside plaster and the outside whitewash is taken as equal to the outside plaster. There is no need for separate calculations for whitewashing and distempering. The number of coats should be mentioned clearly. The surfaces having corrugations are considered flat surfaces, however, there are provisions for increasing the area for compensating the corrugations as follows:

(1) For corrugated iron sheets, the calculated area is multiplied by 1.14.

(2) For the semi-corrugated asbestos cement (AC) sheets, the area is multiplied by 1.10.

(3) For corrugated A.C. sheets having large-sized corrugations, the area is multiplied by 1.20.

1.10.12 Painting

The measurement and payment of the painting work is done in sq m. The area of the doors and windows are taken as flat and girth is not measured. The painting work is done in two or three coats. The paint is done on one coat of the primer. The equivalent plain area is calculated by multiplying some factors given in Table 1.5:

Table 1.5: Multiplying factors

Items	*Multiplying Factor for Both Sides*
Fully glazed	1.0
Partly panelled/partly glazed/flush door	2.0
Venetian door /louvered	3.0
Panelled/framed and braces/ledged/ battened and braced	2.25
Guard bars, gratings, grills, railings	1.0
Nainital pattern roof	2.20
Corrugated iron sheets	2.28

1.10.13 Other Items

There are other miscellaneous items that are not discussed in the previous sections. An estimation engineer should consider the following points too:

(1) The rainwater vents, waste pipes, surface drains etc. are measured in running meters. The length of the pipes is measured along the centreline of the pipe and fittings.

(2) Sanitary fittings like urinals, closet pans, flush pipes, door handles, strap hinges, bolts, etc. are measured and paid in numbers.

(3) There are items like site clearance, site dressing, decoration works, etc. which are difficult to measure. These items are given at some lump sum rate.

(4) The cost of electrification is given as a percentage of the estimated cost of the building. It is taken as 8% of the total estimated cost.

(5) The cost of water supply and sanitary works is also taken as 8% of the total estimated cost of the building.

1.11 Rules for Deduction (As per IS:1200)

There is a need to understand the rules of deductions. Most of the rules for the deduction are discussed in the previous section. However, some of the important rules not discussed previously are discussed in detail.

1.11.1 Opening in Masonry Work

The deductions are made for the opening in the masonry work. For example, the deductions are done for the opening provided for the doors and windows. The deductions for the openings are calculated as follows:

(1) **Rectangular opening:** If the breadth is B, Height is H and width is W, the deduction is calculated by calculating the volume of the masonry to be deducted. The deduction for rectangular opening is calculated as follows:

$$Deduction = Breadth\ (B) \times Height(H) \times Width(W)\ \ cu.m$$

Figure 1.1 (a) shows the deduction for the rectangular opening.

(2) **Door and window opening (with small segmental arches):** In the case of the small segmental arch, the deduction is made for the rectangular part only. The segmental arch is considered as solid portion

which helps in getting extra expenses for making the arch. Deduction for the door and window opening with small segmental arch is given as follows:

$$Deduction = Breadth\ (B) \times Height(H) \times Width(W)\ cu.m$$

The height of the opening is taken up to the springing line only as shown in Figure 1.1(b).

(3) **Door and window opening (with large segmental arches):** The openings of doors and windows with large segmental arches are measured in two parts i.e. area of the rectangular portion and the area of the segmental arch portion. Figure 1.1(c) shows the door and window opening with large segmental arches. The area of the rectangular portion is calculated by multiplying the breadth and height. The calculation for the deduction for such sections is done in the following steps:

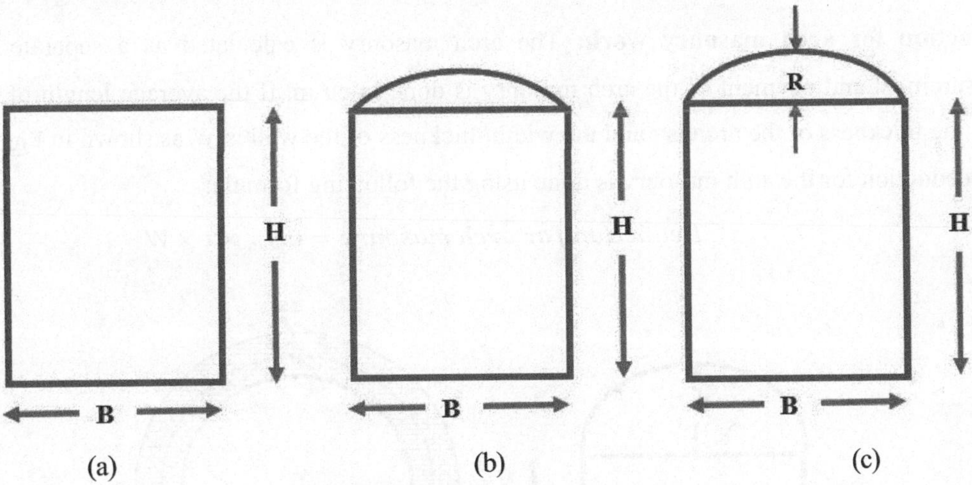

Figure 1.1: Opening for masonry

If R is the rise of the arched portion, B is the breadth, W is the width or thickness of the wall, and H is the height as shown in Figure 1(c).

$$Area\ of\ the\ rectangular\ portion = B \times H$$

$$Area\ of\ the\ large\ segmental\ portion = \frac{2}{3} \times B \times R + \frac{R^3}{2B}$$

The term $R^3/2B$ is neglected for the calculation of the final deduction. So, the total deduction in volumetric calculation for the door and windows opening with large segmental arches is given by the following formula:

$$Deduction\ (with\ large\ segmental\ arch) = (B \times H + \frac{2}{3} \times B \times R) \times W$$

(4) **Doors and windows with semi-circular arches:** If a door or window is to be provided with a semi-circular segmental arch, the area of the rectangle and arch is added to get the deduction. The area of

the semi-circular arch is $\pi D^2/8$, but the exact area for estimation is not calculated. In such cases, the approximate area of the semi-circular arch is used. The deduction for such cases is taken as follows:

$$Area\ of\ the\ rectangular\ portion = B \times H$$

$$Area\ of\ semi-circular\ arch = \frac{3}{4} \times B \times R$$

$$Deduction\ for\ the\ opening\ with\ semi-circular\ arch = \left(B \times H + \frac{3}{4} \times B \times R\right) \times W$$

It is worth noting that the deduction for the opening having elliptical arches is also calculated using the same formula as used in the case of the semi-circular arches. Figure 1.2 (a) shows the opening with a semi-circular arch.

(5) **Deduction for arch masonry work:** The arch masonry is calculated as a separate item. The measurement and payment of the arch masonry is done in cu m. If the average length of the arch is L_{avg}, the thickness of the arch is t and the width/thickness of the wall is W as shown in Figure 1.2 (b). The deduction for the arch masonry is done using the following formula:

$$Deduction\ for\ arch\ masonry = L_{avg} \times t \times W$$

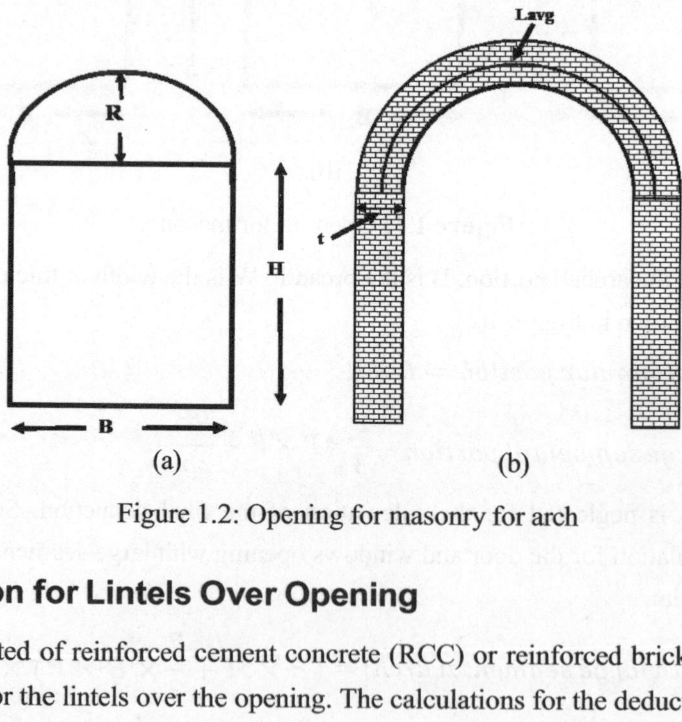

Figure 1.2: Opening for masonry for arch

1.11.2 Deduction for Lintels Over Opening

The lintels are constructed of reinforced cement concrete (RCC) or reinforced brick concrete (RBC). So, the deduction is made for the lintels over the opening. The calculations for the deduction of the lintels over openings are as follows:

$$Length\ of\ the\ lintel\ (L) = B + 2\,t'$$

Here, B is the clear span, and t′ is the bearing as shown in Figure 3. However, if the bearing is not given, it can be taken equal to the thickness of the lintel. The bearing should not be less than 120 mm in any case. Deduction for the lintel is given using the following formula:

$$Deduction\ for\ lintel = L \times t \times W$$

L is the length of the lintel, t is the thickness of the lintel and W is the width of the masonry wall.

1.11.3 Deduction for Plastering

There are a few rules for deduction in plastering. The deductions are the same for openings as discussed in the case of masonry. The rules for the deduction in plastering are listed as follows:

(1) The deduction is not done for the ends of the beams and small holes up to 0.5 sq. m. There is no provision of the addition for jambs, soffits, etc. at the same time.

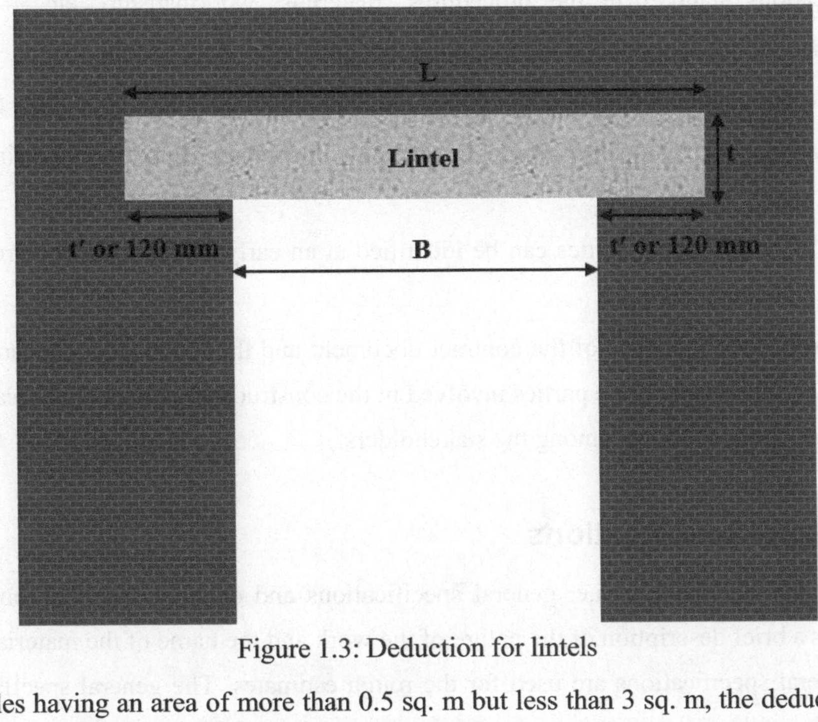

Figure 1.3: Deduction for lintels

(2) For the holes having an area of more than 0.5 sq. m but less than 3 sq. m, the deduction is done only for one face and the other face is left to compensate for the area of jambs, soffits, etc.

(3) In case of the opening having an area of more than 3 sq. m, the deduction is done on both the faces but the jambs, soffits, etc. are calculated separately. The area of jambs, soffits, etc. is added in the plastering separately.

1.12 DESCRIPTION/SPECIFICATIONS OF ITEMS

The specification may be defined as the description of the requirements, dimensions, materials, procedures, etc. that are used to achieve the construction of any building or structure. The specifications help in maintaining the quality of the work. The Public Work Department (PWD) provides the different norms of the work and specifications for the different items of the work. The schedule of rates is a document that consists of the rates applicable to the different items of work. The specifications are given for the following purposes:

(1) Specifications give a clear description of the work. The steps involved in achieving the construction goal are mentioned in the specifications that help in the estimation of costs.

(2) The specifications standardize the procedures, materials, workmanship, etc. It gives consistent materials and products for the different projects.

(3) The detailed specifications help in the calculation of the accurate probable cost of the project.

(4) The specifications help in quality assurance. It helps ensure the quality of the final construction output.

(5) The possible risks and uncertainties can be identified at an early stage of the construction project if clear specifications are provided.

(6) The specifications are the basis of the contract document and they help in understanding the rights, responsibilities, and duties of the parties involved in the construction project. The clear specifications may resolve or reduce disputes among the stakeholders.

1.12.1 Types of Specifications

The specifications are of two types i.e. general specifications and detailed specifications. The **general specification** gives a brief description of the nature of the work and the name of the materials to be used in construction. General specifications are used for the rough estimates. The general specifications are not included in the contract document.

The **detailed specifications** are included in the contract. These are prepared separately in a detailed manner. The detailed specifications describe the type of material, method of preparation of material, proportion of material, workmanship, etc. The specifications of the different items of the work are prepared in the same order in which the work is to be conducted.

1.12.2 Specifications of Items

The detailed specifications are given by the agencies like Public Works Department (PWD). These are prepared separately for the different items of the work. The specifications are discussed for different items of the work in detail.

(1) **Site clearance:** The site should be cleared before the beginning of the cutting and filling. The shrubs, grass, vegetation, and plants up to 30 cm girth measured above 1 m of the ground level. The waste material should be removed up to a distance of 50 m from the boundary of the construction area. The roots of the trees and saplings are removed. The trees with a girth of more than 30 cm at the level of 1 m above ground level, should be cut only after taking permission in writing from the engineer-in-chief. Engineer-in-chief decides the location to dispose off the earth.

(2) **Mode of carriage:** The mode of carriage during earthwork is decided by the engineer-in- chief.

(3) **Lead:** The horizontal distance from the point of excavation to the place where it is transported is known as lead. The lead should be the shortest distance, however, if the shortest distance is not possible due to some obstructions, the shortest practical route is taken. The mode of transportation like animal carriage or mechanical transport should be selected.

(4) **Lift:** It is the vertical distance from the ground level up to which the earth is excavated or deposited. The material excavated up to 1.5 m below the ground level and the material deposited up to 1.5 m above the ground level are included in the earthwork rate.

(5) **Cement:** should conform to the standards of the Indian Standard Institute (ISI). The initial setting time should be at least 30 minutes and the final setting time should not be more than 10 hours. The color of the cement should be grey. It should have the required mechanical properties like compression strength and fineness. The cautions given in the specification book should be followed. The cement bags should be stored carefully in a dry place, having no leakage and no moisture. The flooring of the storage place should be in two layers of bricks on the consolidated earth. The stacking of the cement bags should be done 150 mm to 200 mm above the floor level. The wooden planks can be used for this purpose. The cement bags should be stacked at least 450 mm away from the wall to avoid the penetration of moisture.

(6) **Fine aggregates:** The aggregates that pass through the sieve of size 4.75 mm are called fine aggregates. Natural sand, crushed stones, gravel, broken bricks, etc. can be used as fine aggregates. These should be clear and free from impurities like organic matter, dust, etc. It should be in an

angular shape and should have a fineness modulus of 2.5 to 3.0. The fine aggregates should be an inert material. These shall be stacked so that the dust or any other foreign material does not come in contact.

(7) **Coarse aggregates:** Coarse aggregates should be clear and free from impurities. These should possess required mechanical properties like compression strength, resistance to abrasion, toughness, etc. These should be cubical. The size of the coarse aggregate should be from 15 mm to 40 mm. The artificial aggregates should conform to the provisions of IS 383:2016.

(8) **Fly ash:** The fly ash should be obtained from the nearest power plant. There may be a problem of dust in using the fly ash. So, water is sprinkled over it to keep it moist.

(9) **Water:** Water should be clean and free from impurities like salts, alkalis, oils, salts, sugars, etc. The pH value of the water used in the mixing and preparation of the concrete mix should not be less than 6. The tests for acidity, alkalinity, and dissolved solids should be performed as per IS codes. If the *water is found suitable for mixing, it will be suitable for curing too.*

(10) **Concrete:** The concrete is designated in grades. The concrete is classified into three categories i.e. ordinary concrete, standard concrete and High strength concrete. Table 1.6 shows the designation and classification of concrete.

Table 1.6: Grades of the Concrete

Concrete Group	*Grade Designation*	*Characteristic Compressive Strength (N/mm²)*
Ordinary Concrete	M 10	10
	M 15	15
	M 20	20
Standard Concrete	M 25	25
	M30	30
	M 35	35
	M 40	40
	M 45	45
	M 50	50
	M 55	55
	M 60	60

High Strength Concrete	M 65	65
	M 70	70
	M 75	75
	M 80	80
	M 85	85
	M 90	90
	M 95	95
	M 100	100

The letter M represents the mix and the strength is the characteristic compressive strength in N/mm^2. The compression tests are done on the concrete cube of $150\times150\times150$ mm^3 size and the strength is measured after 28 days.

Concrete having a grade of more than M 60 should be designed according to the parameters from specialized literature of experimentation. In such cases, normal code provisions are not followed. The characteristic strength is the strength of the concrete below which not more than 5% of samples are expected to fall. The minimum grades, maximum water-cement ratio, and minimum cement content are given in Table 1.7. The minimum grade for the plain concrete for mild exposure is not specified. The parameters mentioned in Table 7 are related to the exposure of the mix to different conditions. The lower-grade concrete can be used as lean concrete in the foundation.

Table 1.7: Minimum grade and cement content for weather conditions

Exposure	*Plain Concrete*			*Reinforced Cement Concrete*		
	Minimum cement content (Kg/m^3)	Maximum water cement ratio	Minimum Grade	Minimum cement content (Kg/m^3)	Maximum water cement ratio	Minimum Grade
Mild	220	0.60	NS	300	0.55	M 20
Moderate	240	0.60	M 15	300	0.50	M 25
Severe	250	0.50	M 20	320	0.45	M 30
Very Severe	260	0.45	M 20	340	0.45	M 35
Extreme	280	0.40	M 25	360	0.40	M 40

The workability of the concrete is also an important parameter that needs to be tested. The concrete mix selected for the construction should be sufficiently workable in the placing conditions. It should tend to get properly compacted by the available compaction instruments. IS 1199 is used to understand the workability required in different conditions. Table 1.8 shows the desired workability of concrete in different types of constructions.

Table 1.8: Workability of concrete

Placing Conditions	Degree of Workability	Test Values
Concrete for shallow sections, concrete for pavement built by pavers	Very Low	0.75-0.80 (Compaction factor value)
Mass concrete, slabs, beams, and columns with light reinforcement	Low	25 mm – 75 mm (Slump value)
Slabs, beams, and columns with heavy reinforcement, pavements with hand placement, canal lining, strip footing	Medium	50 mm – 100 mm (Slump value)
Pumped concrete	Medium	75 mm – 100 mm (Slump value)
Filling of the trench	High	100 mm – 150 mm (Slump value)
Termite concrete	Very high	Determined by flow test

There is a need for very strict control in case of very low workability. So, the slump test is not suitable for such conditions. The values for the workability are determined by using the compaction factor test. Similarly, when very high workability is required, it is determined by using the flow test.

(11) **Damp-proof course (DPC):** The thickness of the damp-proof course may be taken 2 cm. The cement and coarse sand in 1: 2 proportion are used for the preparation of mortar. The waterproofing material should be added carefully. The amount of waterproofing material can be taken 1 Kg per bag of cement. For a DPC of 2.5 cm, the cement, sand, and aggregates are taken in the proportion of 1:1.5:3. The ordinary Portland cement is used and the sand should be clean. The aggregates should be inert and free from the impurities. The cement is mixed with waterproofing material and after that, the aggregates are mixed. The curing of DPC is done for 7 days.

(12) **Brickwork:** The bricks should be stored by stacking. These should be stacked at the nearest possible distance from the construction site. It saves the labour's effort in transporting them to the construction work. The stacking of the bricks should be such that the damage to the bricks remains minimal. The different types of bricks should be stacked separately.

If the type of the brick is not specified, the first-class bricks are used. The bricks should be soaked in the water tank for 12 hours before using them in construction. The mortar for the brick masonry should be of cement sand ratio of 1:3 to 1:6 as per specifications. The thickness of the mortar layer should not be more than 6 mm. The joints should be filled with the mortar. The frog of the brick should be kept upwards while in the top layer, the frog is kept in the downward direction. A maximum height of 1 m is kept for brickwork at a time.

The joints and faces of the wall are cleaned after the day's work. The brickwork is cured for 10 days. The specifications for different items of the work are given in the CPWD Specifications. If the students want to learn more about the specification, they can download the CPWD Specifications by scanning the given QR code.

1.13 CONCLUSIONS FROM THE CHAPTER

The subject 'estimating, costing, and valuation' is an important subject. In this chapter, the introduction to estimating and costing is given to the students. The different types of approvals required for the commencement of a construction project are discussed. The standard format of different types of estimates is also given. The checklist that is required by the estimation engineer is presented in tabular form. The methods of measurements and rules for deductions during the preparation of estimates are also discussed in detail. At the end of the chapter, a brief discussion about the specifications and their role in estimation is discussed.

After completion of this chapter, the students would have a basic knowledge of the introduction of procedures required for the preparation of estimates for construction projects. The students would have developed some interest in the subject at this stage. The different types of rough estimates are useful in different conditions of construction projects. It is important to learn the procedure for the preparation of rough estimates for a civil engineer. The next chapter provides detailed information about the rough estimates.

UNIT SUMMARY

- The process of calculating the quantities of the items and the probable cost involved is known as estimation.
- The estimation of probable cost helps the owner in decision-making. In the case of government agencies, probable cost helps in obtaining administrative approval for the execution of the project.
- Administrative approval may be defined as the 'formal acceptance of the proposal' by the competitive authority. At the initial stage, rough estimates and preliminary plans are required for administrative approval.
- The expenditure sanction may be defined as an agreement by the administration/ competitive authority on the proposed cost. Any expenditure before this sanction is not allowed for the proposed construction.
- The formal acceptance of the detailed proposal including designs, detailed estimates, detailed plans, etc. by technical departments is known as 'technical sanction'.
- Approximate estimates are the estimates that give the rough or approximate cost of the construction.
- The preliminary survey data and soil reports with a line diagram of the project can be used for the preparation of approximate estimates.
- The detailed estimate provides an accurate estimate of the probable cost of construction in a detailed manner.
- The detailed estimate is prepared in two steps i.e. Computation of quantities through detailed measurements and Abstract estimate.
- A revised estimate is provided when there is a deviation in the quantity and rate of the item as compared to the estimate provided by the estimator.
- A comparative statement in the abstract format showing the variation of the quantity and rates is provided with the revised estimate.
- If some independent work that is different from the sanctioned work is required, a supplementary estimate is provided.
- Renovation estimates are required for the improvement of the damaged structures.
- Repair and maintenance are performed at a relatively smaller scale; however, the renovation work may include reconstruction of the damaged structures too.

- A checklist of the items helps the estimation engineer to prepare an accurate estimate. It ensures that all the items are included in the estimate and no critical construction component is left.
- A checklist of items gives transparency in the estimation as the stakeholders and clients can review the checklist to understand the preparation of the estimate.
- The measurements of the items and their quantity are presented in the measurement sheet and the cost of the construction is calculated in the abstract sheet.
- The face sheet briefly gives the total cost of the project.
- Indian Standard Code IS 1200 provides information about the different items of work, the unit of measurement, and the mode of measurement.
- The earthwork is measured in cu m and the unit of payment is % cu m.
- The size of a modular brick is 19 cm× 9 cm× 9 cm, and the nominal size of the modular brick is 20 cm × 10 cm × 10cm.
- The steel works and the ironwork are measured and paid by weight generally. Plain rolled sections, steel bars, etc. are measured in Kg or quintals.
- The steel table can be used for obtaining the weight/running m for the different sections.
- The area of the whitewashing and distempering is kept the same as that of plastering and no extra calculation is done.
- The inside whitewashing is taken as equal to the inside plaster and the outside whitewash is taken as equal to the outside plaster.
- The specification may be defined as the description of the requirements, dimensions, materials, procedures, etc. that are used to achieve the construction of any building or structure.
- The specifications help in maintaining the quality of the work. The Public Work Department (PWD) provides the different norms of the work and specifications for the different items of the work.
- The general specification gives a brief description of the nature of the work and the name of the materials to be used in construction.
- The detailed specifications describe the type of material, method of preparation of material, proportion of material, workmanship, etc.

EXERCISES

Multiple Choice Questions

1. Which sanction is required to begin the expenditure on any contract?

   (a) Administrative approval (b) Expenditure sanction

   (c) Technical sanction (d) none of the above

2. If you are assigned a task to prepare an estimate for improving a damaged building, what type of estimate will you prepare?

   (a) Renovation estimate (b) Supplementary estimate

   (c) Repair and maintenance estimate (d) Revised estimate

3. Wall plaster is generally measured in:

   (a) Cu m (b) Running m (c) Sq m (d) None of the above

4. If the depth of the excavation is not more than 30 cm, the excavation is measured in:

   (a) Cu m (b) Running m (c) Sq m (d) None of the above

5. The nominal size of a modular brick is:

   (a) 20cm×10cm×5cm (b) 19cm×9cm×9cm

   (c) 19cm×9cm×4.5cm (d) 20cm×10cm×10cm

6. The earthwork is measured in:

   (a) Cu m (b) % Cu m

   (c) Sq m (d) None of the above

7. The earthwork is paid in:

   (a) Cu m (b) % Cu m

   (c) Sq m (d) None of the above

8. The unit of payment and unit of measurement for the damp proof course (DPC) is:

   (a) Running m (b) Cu m

   (c) Sq m (d) Meter

9. What is the multiplying factor for both sides of the louvered door?

   (a) 2 (b) 3 (c) 3.5 (d) 2.5

10. In the lintel above the door opening, the bearing should not be less than:

    (a) Thickness of wall (b) 100 mm (c) 115 mm (d) 120 mm

11. The slump value for the pumped concrete should be:

(a) 50 mm – 75 mm (b) 75 mm – 100 mm

(c) 50 mm – 100 mm (d) 75 mm – 125 mm

12. The depth of excavation up to which the rate of excavated or deposited material is included in the earthwork:

(a) 1.5 m (b) 2.0 m (c) 2.5 m (d) 3.0 m

Answers of Multiple-Choice Questions

1. (b), 2. (a), 3. (c), 4. (c), 5. (d), 6. (a), 7. (b), 8. (c), 9. (b) 10. (d) 11. (b) 12. (a)

Short and Long Answer Type Questions

1. Being a civil engineer, what type of approvals and sanctions will you take for a construction project? Explain in detail.
2. Explain the responsibilities of the estimation engineer at different stages of a construction project.
3. State the difference between the repair and maintenance estimate and the renovation estimate.
4. What is the purpose of the face sheet? Draw its standard format.
5. Enlist the provisions of measurement of RCC work.
6. Explain the rules of deduction for the opening with a large segmental arch.
7. Discuss the rules of the deduction for the lintel over the door opening.
8. What are the different types of specifications? Explain briefly.
9. Explain the different specifications for site clearance.
10. According to the specifications of the CPWD, what are the characteristics of the suitable water for mixing and curing of concrete?
11. Explain the concept of deduction of arched masonry with a neat sketch.

KNOW MORE

IS 1200 is an Indian Standard Code that provides the norms for the estimating and costing of the different items of work during construction. IS 1200 provides the methods of measurement of excavation, concreting, painting, plastering, flooring, etc. The students who are interested in learning the subject of estimating and costing must go through these codes. Interested students can scan the given QR code to download the different parts of IS 1200.

REFERENCES AND SUGGESTED READINGS

1. Datta, B. N. *Estimating and Costing in Civil Engineering*, 1st Edition UBS Publisher Distributors Pvt. Ltd, New Delhi

2. Chakraborti, *M. Estimating, Costing, Specification & Valuation in Civil Engineering,* 1st Edition, Monojit Chakraborti, Kolkata

3. Peurifoy,R. L. & Oberlender, D. G. *Estimating Construction Costs*, 5th Edition, McGraw Hill Education, New Delhi

4. Singh, G. & Singh, J. *Estimating Costing and Valuation*, 1st Edition, Standard Publishers Distributers, Delhi

5. IS 1200 (All Parts), *Indian Standard, Method of Measurement of Building and Civil Engineering Works,* Bureau of Indian Standards (BIS), New Delhi

6. Government of India, *Central Public Works Department Specifications*, Vol 1, 2019, published under authority of Director General, CPWD, New Delhi, printed and distributed by Jain Book Agency (Central), New Delhi

2 Approximate Estimates

UNIT SPECIFICS

Through this unit we will discuss the following aspects:

- Definition and purpose of the approximate costs
- The service unit method of approximate estimation
- Approximate estimation by plinth area rate technique
- Cubical content method and typical bay method
- Approximate estimate for roads, Railways, and bridges/culvert

To enhance the student's practical understanding, the different types of approximate estimates are discussed with numerical problems. Some of the important information is shared using the solved problems for a better understanding of the concepts.

This unit consists of exercises with long and short-term questions, multiple-choice questions, a list of references, and suggested reading to make the students learn through practice. Some extra information that may interest the students is provided in the form of QR codes. The students can scan these QR codes to reach the source of further knowledge on the different topics.

RATIONALE

This chapter on the approximate costs presents the different techniques for the calculation of the rough costs. The techniques for the calculation of the rough cost of the construction of the buildings, bridges, highways, etc. are discussed in detail. The approximate cost based on the plinth area rate and cubic content rate are presented in detail. Some solved numerical problems on different techniques of rough estimates are given for a better understanding of the students. The content covered in this unit equips the students with procedures of calculating the approximate costs.

The knowledge of preparation of the approximate estimates or rough estimates is very important for civil engineering students and professionals. It helps the students understand and calculate the approximate probable cost of the construction projects in a short period. The concept of plinth area estimate, cubical content technique, and methods for the calculation of approximate estimates are essential to understand for a civil engineering student. The knowledge gained through this chapter will help the students in their civil engineering career and they will be able to calculate the rough probable cost of a construction project quickly and without the detailed drawings and plans.

PRE REQUISITES

The introductory knowledge of the estimates and their types is required for understanding this chapter.

UNIT OUTCOMES

The list of outcomes of this unit is as follows:

U2-O1: Understand the different methods of approximate estimates.

U2-O2: Prepare the approximate estimate of civil engineering structures.

U2-O3: Select a suitable technique for an approximate estimate in different conditions.

Unit-2 Outcomes	EXPECTED MAPPING WITH COURSE OUTCOMES (1- Weak Correlation; 2- Medium correlation; 3- Strong Correlation)				
	CO-1	CO-2	CO-3	CO-4	CO-5
U2-O1	1	3	1	-	-
U2-O2	-	3	2	-	-
U2-O3	-	3	-	-	-

2.1 Approximate Estimate and its Purpose

The students are familiar with the different types of estimates and their formats at this stage. As discussed in the previous chapter, there are two types of estimates i.e. approximate or rough estimate and detailed estimate. The preparation of detailed estimates requires more time and resources than rough estimates. So, approximate estimates are prepared in the very beginning and presented for getting administrative approval. To know more about the approximate estimate, scan the QR code.

The approximate estimate may be defined as a primary calculation of the cost of a construction project using some preliminary information. The approximate estimate is prepared for many reasons. The purposes of the approximate estimate are as follows:

(1) The approximate estimate helps in understanding the feasibility of the proposed construction project. It answers the question of whether the project is viable under given circumstances.

(2) The approximate estimate provides a rough idea about budget planning. The availability of funds and the need to secure the funds are decided based on the approximate estimate.

(3) The rough estimate guides the decision-makers and managers in deciding on the acceptance or rejection of the proposed project.

(4) A rough estimate is required to get the administrative approval of the projects. The rough estimate is attached with the application for administrative approval.

(5) The approximate estimate gives an idea of the risks involved in the construction projects. If the risks of the construction project are understood at the early stage, the action plan can be prepared soon.

2.2 Methods of Approximate Estimate

The approximate or rough estimate is useful for decision-makers, administrators, and planners. Detailed drawings or plans are not required for the preparation of approximate estimates. This type of estimate saves time and cost and provides a foundation for the initial planning of the project. The approximate estimate is prepared using experience on similar projects. The approximate estimate is prepared using the following methods:

(1) Service unit method

(2) Plinth area rate method

(3) Cubical content method

(4) Typical bay method

(5) Approximate quantity method

(6) Empirical Equations

2.2.1 Service Unit Method

The service unit method is also known as the unit rate method of approximate estimate. In-this technique, the cost of a unit is considered. The total cost of the project is calculated by multiplying the cost per unit by the number of units in a structure. The cost per unit for the highways is considered as cost per km while for the bridges the cost per m span is considered.

For some buildings like hospitals, cost per bed or for theatres cost per seat is calculated. The service unit method provides a rough idea of the project cost based on the previously built structure of the same configuration and the same location. The following precautions are taken while calculating the approximate cost using the service unit method:

(1) The rates of similarly designed and constructed projects in a similar locality are required for the implementation of the service unit method. The rates may vary with the time.

(2) The price level of the construction of units may vary with time and the location of the proposed project. So, the variation of the price level must be considered while using the service unit method.

(3) There may be some changes in the specifications of the proposed project that will impact the final cost of the project. The variation of the specification should be considered by the estimation engineer while using this technique.

(4) The condition of the soil changes when the location of the project is varied. So, the estimation engineer should understand the condition of soil, topography characteristics, and drainage conditions of the proposed location.

(5) The cost of the construction varies with the number of units to be constructed. In general, the per unit cost reduces when the number of units to be constructed is relatively more in number. For example, the per unit cost of a 50-bed hospital will be less than the per unit cost for a 20- bed hospital.

2.2.2 Plinth Area Method

The plinth area method is an important method for approximate estimation. In this technique, the plinth area of the building is calculated and the cost per sq m of the plinth area is considered for calculating the total cost. The total cost is calculated by multiplying the cost per sq m of plinth area for a similar building in the same locality by the plinth area of the building. The estimation engineer should be familiar with the following important terms for using the plinth area method:

(1) **Plinth area:** The plinth area may be defined as the total covered area of a floor including the thickness of the walls. The plinth area is the area enclosed by the outer walls. It is calculated by the external dimensions of the building. The plinth area and other concepts can be understood by scanning the QR code.

(2) **Carpet area:** The carpet area may be defined as the total usable area of the building. The total carpet area is calculated by subtracting the area of the external walls from the plinth area. However, the area of balconies, corridors, porches, staircases, etc. is not included. Figure 2.1 clears the concept

of the carpet area and plinth area. Indian Standard Code IS 3861: 2002, give the details of the measurement of the plinth area. The following areas are included in the plinth area:

(a) The area of the walls at the floor level is considered in the plinth area.

(b) If the columns are projected beyond the cladding, the plinth area is calculated using the external face of the cladding.

(c) The 100 % of the area of the parapet and bacony protected from the above is considered in the plinth area while for the unprotected case, 50% of the area is considered.

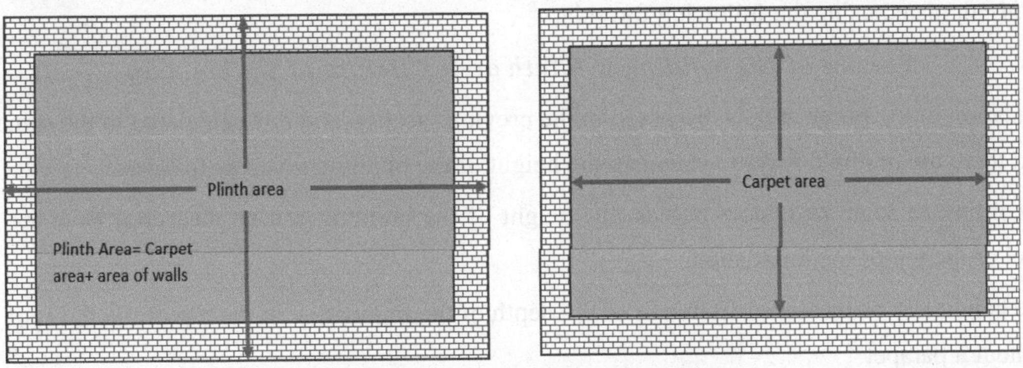

Figure 2.1: Plinth area and carpet area

However, the area of the mumty, loft, architectural band, cornice, box louver, etc. are not considered while calculating the plinth area.

(3) **Floor area:** The floor area may be defined as the plinth area excluding the walls. The floor area includes the area of all the rooms, kitchen, verandah, porch, water closets, etc. The area of the walls includes the openings in the wall too.

(4) **Circulation area:** The circulation area is the area that is required for the movement of the residents to the rooms and other places. The circulation area is a free space that is given in the houses in the form of verandahs, corridors, balconies, lift shafts, etc.

The plinth area can be measured from the line plan at the beginning of the project. A line plan of the different facilities like rooms, kitchen, closet, etc. is prepared according to the requirement of the owner. The floor area is calculated in the first step. The carpet area for the rooms and the circulation area (area of the porch, corridors, kitchen, water closet, etc.) are added to get the floor area. The floor area is added to the area of walls to get the plinth area. The plinth area can be calculated from the carpet area too. The plinth area from the carpet area is calculated using the following equation:

$$Plinth\ area = Carpet\ area + Area\ of\ walls + Circulation\ area$$

The area of walls is taken 15% to 17% of the carpet area. However, in the case of reinforced cement concrete (RCC) framed structures, the area of walls is taken 8% of the carpet area. The circulation area is taken 10% to 15% of the carpet area. It is important to note that the area of walls depends on the shape of the structure.

2.2.3 Cubical Content Method

In the cubical content method, the volume of the building is calculated. The rate per unit volume of the building is multiplied by the volume of the building to get the cost of the building. The volume of the building is calculated using the following formula:

$$Volume\ of\ the\ building = Plinth\ area \times Height\ of\ the\ building$$

The calculation of the plinth area is discussed in the previous section and it is calculated in the same way. However, there are multiple ways to calculate the height of the building given as follows:

(1) According to some regulatory bodies, the height of the building can be measured from the ground level or the top of the foundation.

(2) The height can be measured from half of the depth of the foundation to the top of the flat roof without a parapet.

(3) The height is measured from the half of the rise of the pitched roof or parapet in the flat roof. This criterion is generally used in India.

(4) In the case of a multi-storey building, the height from the floor of one storey to the floor of the upper storey is measured.

2.2.4 Typical Bay Method

The typical bay method is used where the pattern of facilities to be constructed is repeated in a construction project. For example, there are multiple platforms on a railway station. So, in such cases, the typical bay technique can be used. A bay may be defined as the distance between two supports like columns or beams. The cost of the construction of a bay is determined based on experience and multiplied by the number of bays to be constructed to get the approximate cost of the project. The probable cost in the typical bay method is calculated using the following formula:

$$Approximate\ cost = Cost\ per\ bay \times Number\ of\ bays$$

Changes in the rates and the location of the project are also considered while implementing this technique. The extra cost of end walls and other items other than the bay are also considered for the calculation of the accurate approximate cost.

2.2.5 Approximate Quantity Method

In the approximate quantity method, the cost of the construction is calculated by multiplying the cost per unit running m of the wall by the total length of the walls. The plan of the building can be used for the calculation of the length of the walls. The different sections of the walls are considered separately and the calculations from the foundation to the roof are done. The cost per running m of the wall consists of the cost of excavation, foundation construction, brickwork, concreting, roofing, etc. is included. The following points should be considered in the approximate quantity method:

(1) The items should be grouped and the rate should be fixed based on the experience of the estimator.

(2) The calculation of cost per running m from the foundation to the plinth level can be calculated and similarly, the cost of the superstructure from the plinth level to the roof can be calculated separately. The cost of woodwork, flooring, decoration, etc. is also added to get a more accurate estimate.

(3) There is the provision of giving the allowance for the items that may be missed during the calculation.

The approximate quantity method is a little bit time-consuming, however, it gives a more accurate estimate as compared to the other approximate estimate techniques. The rates considered in the approximate quantity method should be realistic and updated to get the approximate cost near to the detailed cost.

2.2.6 Empirical Equations

Central Building Research Institute (CBRI), Roorkee, India prepared some empirical equations for the calculation of different items based on the plinth area. These equations are used to calculate the approximate materials and labor required for a construction project. The approximate cost of the construction project is calculated using the current rate of different items and labor. Scan the QR code to understand the CBRI equations through a video.

The water supply charges are added at the rate of 5% of the cost calculated. This cost is increased by 10% for considering the profit of the contractor. If P_a is the plinth area, different items of the work are calculated as follows:

(1) $Bricks\ (in\ 1000) = 0.226 \times P_a + 6.680$

(2) $Cement\ (in\ tonnes) = 0.153 \times P_a + 0.57$

(3) $Lime\ (in\ Kg) = 14.5 \times P_a - 35$

(4) $Sand\ (in\ Cu\ m) = 0.47 \times P_a - 7$

(5) $Masons\ (in\ numbers) = 1.335 \times P_a + 28$

(6) $Unskilled\ Helpers\ (in\ numbers) = 4.766 \times P_a - 32$

The plinth area is calculated in the same way as discussed in the previous sections.

2.3 APPROXIMATE ESTIMATE FOR ROADS

The approximate estimate for the roads is prepared according to the cost per km construction of the highway. The cost of the highway depends on the thickness of different layers, the number of lanes, the length of highways, and the facilities provided by the highways. The cost of acquiring the land also varies significantly with location. So, the cost of land acquisition is considered according to the location and current prices. The approximate estimate for the roads is prepared by considering the following points:

(1) **Land acquisition:** The cost of land and the cost of acquiring the land is considered while calculating the cost of the approximate estimate. The cost of the land varies with the alignment of the highway. The cost of land acquisition is usually very high.

(2) **Cost of excavation and embankment:** The cost of excavation depends on the topographical characteristics of the area. The type of soil to be excavated, the lithology of the region, and the slope associated with the proposed alignment of the land affect the cost of excavation and embankment.

(3) **Cost of layers:** The cost of the layers of the pavement depends on the type of highway to be constructed. The pavement is constructed in the layers of sub-base, base, and surface. The cost of pavement depends on the thickness of these layers. The cost of the highway is calculated by multiplying the cost per km by the length of the highway.

2.4 APPROXIMATE ESTIMATE FOR RAILWAY PROJECTS

The approximate estimate for the railway project is also calculated using the cost per km like highways. The cost of land acquisition is an important factor that impact the cost of the railway projects. The cost per unit for a similar railway line with the same gauge and ballast is calculated and multiplied by the length of the proposed railway line to get the final cost of the project. The major items to be considered while preparing the approximate estimate for the railway projects are enlisted as follows:

(1) Foundation of the railway line and the facilities to be provided

(2) Cutting and filling works

(3) Bridges and their type

(4) Office building

DIPLOMA COURSE BOOKS

S.No.	ISBN	Book Name	Author
37	9789391505516	Engineering Workshop Practice	A. K. Sarathe
38	9789391505653	Environmental Science	Subrat Roy
39	9789355384966	Estimating, Costing and valuation	Sandeep Panchal
40	9789355389510	Fluid Mechanics & Hydraulic Machinery	Suman Chakraborty, Sourav Mitra, Aditya Bandopadhyay
41	9789391505592	Fundamentals of Electrical and Electronics Engineering (with Lab Manual)	Susan S. Mathew, Saji T. Chacko
42	9789355387240	Fundamentals of Power Electronics	Soumitra Kumar Mandal
43	9789355389602	Geotechnical Engineering (Theory & Practicals)	Neelima Satyam
44	9789355386137	Induction, Synchronous, and Special Electrical Machines	K. Selvajyothi, S. Tamilselvi, Sarat Kumar Sahoo
45	9789355382290	Industrial Engineering & Management	A. Manimaran
46	9789355384959	Internet of Things	Sujata Pal
47	9789355388650	Introduction to DBMS: Theory & Practicals	Myneni Madhu Bala
48	9789355384294	Introduction to e-Governance	Kamal Kant Verma
49	9789355386564	Introduction to Electric Generation Systems	Lalit Chandra Saikia, Nalin Behari, Deb Choudhury
50	9789391505585	Introduction to IT Systems (with Lab Manual)	Prashant Joshi
51	9789355382900	Linear Integrated Circuit	B. T. Krishna
52	9789355388148	Manufacturing Engineering	Santosh Kumar
53	9789355383877	Material Science & Engineering	Vinod Kumar, Nisheeth k. Prasad
54	9789391505424	Mathematics-I	Deepak Singh
55	9789391505523	Mathematics-II	Garima Singh
56	9789355385994	Measurements and Metrology	A. Gnanavelbabu, M. Santhana kumar, R. Saranraj
57	9789355388131	Mechanics of Materials and their Testing	R. Velmurugan, Pragathi
58	9789355385048	Microcontroller Applications	Ritula Thakur
59	9789355385062	Microcontrollers and Applications	Santanu Chattopadhyay
60	9789355388100	Mobile and Wireless Communication	Maulin Joshi, Urvashi P. Shukla
61	9789355387790	Production & Operations Management	Mayank Kumar Agrawal
62	9789355388704	Public Health Engineering	Rinku Kumar
63	9789355387455	Python Programming	Rupesh Nasre
64	9789355387189	Renewable Energy Technologies	Sanjay Agrawal, Rajeev K. Mishra
65	9789355387479	Software Engineering	Salman Abdul Moiz
66	9789355389749	Theory of Machines & Mechanisms	G. C. Mohan Kumar
67	9789355387332	Theory of Structures	Arunachalam Subramanian Balu
68	9789355389015	Thermal Engineering - I	V. Raghavan
69	9789355389831	Thermal Engineering - II	Parvesh Antil
70	9789355386779	Transportation Engineering (Theory & Practice)	A.K. Raji, K.K. Babu
71	9789355384287	Water Resource Engineering (Theory & Practice)	Prakash C. Swain, Dillip K. Ghose
72	9789355383051	Web Technologies: Theory & Practical	Raman Dugyala

AICTE'S LIST OF PRESCRIBED TEXTBOOKS

As Per Model Curriculum & Outcome Based Education

aligned with

NATIONAL EDUCATION POLICY 2020

UNDERGRADUATE COURSE BOOKS

S.No.	ISBN	Book Name	Author
1	9789355382160	Advanced Electrical Machines: Theory and Practicals	Bharat Singh Rajpurohit
2	9789355382139	Analog Electronic Circuits: Theory and Practical	Venu Sonti, Sachin Jain
3	9789355383808	Analog Electronic Devices: Theory and Practicals	N. B Balamurugan
4	9789391505158	Basic Electrical Engineering (with Lab Manual)	S.K. Sahdev
5	9789355388087	Basic of Electronics	Brijesh Iyer
6	9789355387127	Basics of Electrical Machines: Theory and Practicals	D. K. Palwalia, U. K. Kalla, R. K. Kumawat
7	9789355389381	Basics of Mechanics	Shubhankar Bhowmick, N. Venkata Swamy Naidu
8	9789355386250	Basics of Thermodynamics	Parmod Kumar, Atul Dhar
9	9789355386984	Biology for Engineers	Annamma Odaneth
10	9789391505141	Chemistry-I	Manisha Agarwal
11	9789355384843	Civil Engineering Societal and Global Impact	Shakuntala Acharya
12	9789355382184	Computer Aided Civil Engineering Drawing	D. Chandra Mouli, B. Swathi
13	9789355382580	Computer Aided Design & Analysis	Hari Kumar Voruganti
14	9789355387592	Computer Architecture	D. Vaithiyanathan
15	9789355385093	Computer Networks	Hari Prabhat Gupta
16	9789355385307	Construction Engineering & Management	Sparsh Johari
17	9789355388728	Control Systems	S.N. Deepa
18	9789355385420	Data Structure and Algorithms	Ranbir Singh Sanasam
19	9789355386700	Design and Analysis of Algorithms	Hari P. Gupta, Rahul Mishra
20	9789355388872	Digital Electronics and System	Abhishek Bhatt
21	9789355386571	Digital Signal Processing	Sanjeev Sharma
22	9789355383396	Disaster management	Pradeep K. Goyal, Anil K. Gupta
23	9789355385222	Discrete Mathematical Structures	Narendra S. Chaudhari, Sharmila S. P.
24	9789355387868	Effective Technical Communication	Pritee Sharma, Sanchita Verma
25	9789355383389	Electrical Circuit Analysis and Network Theory	Shouri Chatterjee
26	9789355388957	Electromagnetic Waves	Uttam M. Pal, Srijith K.
27	9789355384652	Electronic Communication	Vimal Bhatia
28	9789355388711	Embedded Systems	Santanu Chattopadhyay
29	9789355385147	Energy Science and Engineering	M. Rizwan, Majid Jamil
30	9789355387929	Engineering Economics, Estimation & Costing	Yogesh C. Joshi
31	9789355384669	Engineering Geology	R. N. Khare, Ketan Chourasia
32	9789391505066	Engineering Graphics & Design	Pradeep Jain
33	9789391505097	English (with Lab Manual)	Kulbhushan Kumar
34	9789355385963	Environmental Engineering	Chandra Sekhar Matli
35	9789355388117	Fluid Mechanics	Shreenivas N. Londhe, Pradnya R. Dixit
36	9789355388377	Fluid Mechanics & Fluid Machines	R.P. Saini
37	9789355387103	Geotechnical Engineering	Paramita Bhattacharya

(5) Workshop

(6) Cloakroom

(7) Platforms

The other miscellaneous items like sanitary arrangements, drinking water facilities, toilets and washrooms, waiting rooms, etc.

2.5 APPROXIMATE ESTIMATE FOR BRIDGES AND CULVERTS

The approximate estimate of the bridges is calculated according to the cost per unit of span. The following features should be kept in mind while calculating the approximate cost for the bridges:

(1) The cost of the bridges and culverts depends on different factors like the type of bridge, type of foundation, excavation equipment, road type, and the different materials used in the components of the bridges.

(2) The cost of the bridges or the culverts is calculated in the two parts. The cost of the sub- structure and superstructure is calculated separately.

(3) The study on the hydrological data and flood level should be conducted in case the bridge is constructed on the water channels.

(4) The total cost of the bridge is calculated by multiplying the cost per unit span to the span of the bridge.

2.6 APPROXIMATE ESTIMATE FOR IRRIGATION PROJECT

The approximate estimate for the irrigation projects depends on the technique to be used for the irrigation. However, the approximate estimate for the irrigation projects requires a more detailed approach as it involves different hydrological, geological, and agricultural studies. The approximate estimate for the irrigation project is prepared in the following steps:

(1) Aim/objectives of the irrigation project: The objective of the construction of the irrigation project should be understood clearly while preparing the approximate estimate. The irrigation projects may be single-purpose or multi-purpose.

(2) Field data: The different types of field data are required for preparing the approximate cost of an irrigation project. The geological test data and the characteristics of the layers of embedded soil are required for the calculation of the cost of excavation and construction. Similarly, the hydrological data including the level of water and their variation is also required. So, different types of technical survey data are required while preparing the approximate estimate for the irrigation projects.

(3) **Demand forecasting:** The irrigation projects are designed to meet the requirements for a long time So, the forecasting of irrigation water demand for the upcoming 40-50 years is done and the fina water demand is calculated for designing the irrigation system.

(4) **Project formulation and evaluation:** The project is formulated based on the data obtained in the previous sections. The different alternatives of the irrigation projects for achieving the objectives are obtained. The cost of each alternative is calculated. The cost of the irrigation project can be calculated by multiplying the cost per km to the total km. However, the cost can also be calculated by multiplying the average cost per hectare commanded area by the total commanded area. The overhead percentage and contingency cost are taken at 10% each. The cost of land acquisition is also added to the cost of construction. In the absence of the exact cost of land acquisition, it can be taken 12% of the construction cost.

2.7 APPROXIMATE ESTIMATE FOR WATER SUPPLY PROJECT

The preparation of the approximate estimate for water supply projects also requires a detailed approach to irrigation projects. The cost of such projects depends on multiple factors and the estimation of their cost is a complex task. The cost of the water supply project can be calculated in two ways as discussed below:

(1) If the cost of construction of a similar type of water supply system is known, the cost per capita of the water supply is calculated.

(2) The cost per capita of the water supply is multiplied by the number of people for which the water supply system is to be provided to find the cost of the water supply project.

(3) If the data given in point 1 is not available, the water supply system is divided into different units and the cost of each unit is calculated. The cost of land acquisition is calculated at the current rates.

(4) The cost of headwork is calculated by multiplying the cost per m^3 capacity by the total capacity. The cost of pumping machinery is calculated by multiplying the cost per horsepower (HP) by the total power required for the pumping machinery.

(5) The cost of the rising main pipe is calculated by multiplying the cost per running m by the total height of the rising main pipe.

(6) The cost for the treatment units and reservoir is calculated according to the capacity in m^3 and the cost of the distribution system is calculated in running m.

(7) The overall cost is calculated by adding the different costs calculated in the previous steps. The total cost of a water supply system may be given using the following equation:

Approximate cost of water supply (C)

$= Cost\ of\ land\ acquisition\ (C_l) + Cost\ of\ intake\ work\ (C_i)$

$+ Cost\ of\ pumping\ machinery\ (C_p) + Cost\ of\ rising\ main\ (C_r)$

$+ Cost\ of\ treatment\ unit\ (C_t) + Cost\ of\ treatment\ unit\ (C_{tu})$

$+ Cost\ of\ ground\ reservoir\ (C_{gr}) + Cost\ of\ distribution\ system\ (C_d)$

$+ Other\ costs\ (C_o)$

The demand for the water supply should be forecasted accurately as it impacts all the factors that affect the cost of the water supply system.

2.8 CONCLUSIONS FROM THE CHAPTER

At the initial stage of a project, there is a need to provide quick information about the probable cost of the construction. So, the approximate or rough estimates are very important to understand for civil engineering students and professionals. The approximate estimates help the owner in making the decisions and understanding the financial constraints involved in the projects. In this unit, the techniques for preparing the approximate estimates are discussed in detail with numerical examples. Methods like the plinth area method, cubic content technique, service unit method, typical bay method, etc. are discussed in this chapter. The details of approximate estimates for bridges, water supply projects, and irrigation projects are also provided.

After the completion of this chapter, the students would have understood the procedure of approximate estimation. They would have gained expertise in the preparation of rough estimates for the buildings and other types of structures. However, the approximate estimate doesn't give a clear picture of the total cost of the project. So, the approximate estimates can't be used at the advanced stages of the project. The detailed estimates are used for the calculation of the accurate probable cost of construction. A civil engineer or quantity surveyor should have a thorough knowledge of the detailed estimates and their procedures. The next chapter provides detailed information about the detailed estimates.

SOLVED NUMERICAL EXAMPLES

Problem 1: Prepare the approximate estimate for a building with the following data:

- Plinth area : 120 sq m
- Plinth area rate : 60000 ₹/ sq m

- Cost of water supply : 8 % of the cost of the building
- Sanitary and electricity charges : 8.5 % of the cost of the building
- Cost of decoration : 2 % of the cost of the building

The contingencies are taken 6% and supervision charges may be taken 8.5%. Calculate the overall cost of the building project.

Solution:

The plinth area and cost per unit plinth area are given. The overall cost can be calculated as follows:

- Plinth area= 120 sq m
- Rate of plinth area= 60000 ₹/ sq m
- Cost of the building= 120 × 60000 = 7200000 ₹
- Cost of water supply= 8 % of the cost of the building = $7200000 \times \frac{8}{100}$ = 576000 ₹
- Cost of sanitary and electricity= 8.5 % of the cost of the building = $7200000 \times \frac{8.5}{100}$ = 612000 ₹
- Cost of decoration= 2 % of the cost of the building = $7200000 \times \frac{2}{100}$ = 144000 ₹
- Total cost of the building= 7200000+ 576000+ 612000+144000= 8532000 ₹
- Cost of contegencies= 6 % of the total cost of the building = $8532000 \times \frac{6}{100}$ =511920 ₹
- Supervision charges= 8.5% of the total cost of the building= $8532000 \times \frac{8.5}{100}$ = 725220 ₹
- Overall cost of the building project = Total cost of the building+ Contegencies cost+ Supervision charges

  = 8532000+ 511920+ 761940 = **9805960 ₹**

  So, the overall approximate cost for the proposed building project is **9805960 ₹.**

Important note: *It is important to note that the contingencies and supervision costs are calculated for the total cost of the building including the cost of water supply, electricity, sanitary, decoration, and contingencies.*

Problem 2: Prepare an approximate estimate of a hospital with 120 beds. The following details are available for a similar project:

- Carpet area required per bed : 6 sq m
- Area of walls : 20% of the plinth area

- Area of corridor, verandah, etc. : 25% of the plinth area
- Plinth area rate : 120000 ₹/ sq m
- Cost of water supply : 10 % of the cost of the building
- Sanitary and electricity charges : 12 % of the cost of the building
- Cost of decoration : 2.5 % of the cost of the building
- Contingencies : 6% of the total cost
- Work charge establishment cost : 5% of the total cost

Solution:

The carpet area, area of walls, area of corridor, verandah, etc. is given. The plinth area can be calculated from these details:

- Carpet area per bed = 6 sq m
- Capet area for 120 beds = Carpet area per bed× total beds = 6×120 sq m = 720 sq m

Let's assume, P is the plinth area and C is the carpet area. So, the plinth area can be represented as follows:

Plinth area (P) = Carpet area (C) + Area of walls+ Area of verandah, corridor, etc. Area of walls = 20% of the plinth area = 0.20×P

Area of corridor, verandah, etc. = 25% of the plinth area = 0.25×P So, Plinth area (P) = C + 0.20×P + 0.25× P

Solve the equation and calculate the value of P. P = C+ 0.45 P

P- 0.45 P = C

0.55 P = C

P = C/0.55

So, the Plinth area (P) = Carpet area÷ 0.55= $\frac{720}{0.55}$ =1309.09 sq m ≈ 1310 sq m (Approximately)

- Plinth area rate = 120000 ₹/ sq m (Given)
- Cost of the building = Plinth area rate × Plinth area rate = 120000× 1310 = 157200000 ₹
- Cost of water supply = 10 % of the cost of the building = 157200000 × 10/100 = 15720000 ₹
- Cost of sanitary and electricity = 12 % of the cost of the building = 157200000 × 12/100 = 18864000 ₹
- Cost of decoration = 2.5 % of the cost of the building = 157200000 × 2.5/100 = 3930000 ₹

- Total cost of building= 157200000+ 15720000+ 18864000+ 3930000 = **195714000 ₹**
- Cost of contegencies= 6 % of the total cost of the building= $195714000 \times 6/100$ =11742840₹
- Work establishment charges = 5% of the total cost of the building
  = $195714000 \times 5/100$ = **9785700 ₹**
- Overall cost = Total building cost+ Contegencies cost+ Work charge establishment charges
  = 195714000 + 11742840+ 9785700 = **217242540 ₹**

  So, the overall approximate cost of the proposed hospital is **217242540 ₹.**

Important note: *It is important to note that the area of walls and area of circulation area like verandahs, corridors, etc. is given as a percentage of the plinth area, not of the carpet area. The cost of work-charge establishment or supervision charges is calculated for the total cost of the building after adding the water supply charges, sanitary and electricity charges, and cost of decoration to the cost of the building.*

Problem 3: Consider the following data for a building to be constructed :

- Plinth area= 150 sq m
- Height of the building from the ground floor= 4 m
- Height of parapet wall= 0.9 m
- Rate of per cu m volume of construction = 2500 ₹/ cu m Calculate the approximate cost of the building.

Solution:

It can be observed from the problem statement that the rate per cu m volume of construction is given. So, the method to solve this problem will be the cubical content method. The approximate cost for the proposed building is calculated using the following steps:

- Plith area (P) = 150 sq m (Given)
- Height of building from ground level (H_b)= 4 m (Given)
- Height of parapet wall (h)= 0.9 m (Given)
- Height for volume calculation (H)= $H_b + h/2 = 4 + 0.9/2$ = 4.45 m
- Volume of the similar type of building = Plith area (P)× Height for volume calculation (H)
  = 150× 4.45 = 667.5 m
- Rate of per cu m volume of construction = 2500 ₹/ cu m (Given)
- Cost of the building = Rate of per cu m volume × Volume of the similar type of building
  = 2500 × 667.5 = **1668750 ₹**

So, the overall approximate cost of the proposed building is **1668750 ₹.**

Important note: *It is important to note that the height for the volume calculation is calculated by adding the height of the roof from the ground level and half of the height of the parapet. These norms for the calculation of height can be seen in section 2.2.3 of this chapter.*

Problem 4: The construction cost per unit m for a bridge is 900000 ₹. Calculate the approximate cost of construction of a two-lane bridge having a length of 200 m using the Unit area method.

Solution:

In the unit area method, the cost of construction is calculated by multiplying the cost per unit span to the total span of the proposed bridge.

- Cost per unit m= 900000 ₹
- Span of the bridge= 200 m
- Total cost= Cost per unit span× Span of the bridge = 900000×200 = **180000000 ₹**

So, according to the unit rate method, the cost of construction of a two-lane bridge with a 200 m span is **180000000 ₹.**

UNIT SUMMARY

- The approximate estimate may be defined as a primary calculation of the cost of a construction project using some preliminary information.
- The approximate estimate helps in understanding the feasibility of the proposed construction project.
- In the service unit method, the total cost of the project is calculated by multiplying the cost per unit by the number of units in a structure.
- The price level of the construction of units may vary with time and the location of the proposed project.
- The plinth area may be defined as the total covered area of a floor including the thickness of the walls. The plinth area is the area enclosed by the outer walls.
- It is calculated by the external dimensions of the building.
- The floor area includes the area of all the rooms, kitchen, verandah, porch, water closets, etc.
- The circulation area is a free space that is given in the houses in the form of verandahs, corridors, balconies, lift shafts, etc.
- In the cubical content method, the volume of the building is calculated. The rate per unit volume of the building is multiplied by the volume of the building to get the cost of the building.

- The cost of the construction of a bay is determined based on experience and multiplied by the number of bays to be constructed to get the approximate cost of the project.
- In the approximate quantity method, the cost of the construction is calculated by multiplying the cost per unit running m of the wall by the total length of the walls.
- Central Building Research Institute (CBRI), Roorkee, India prepared some empirical equations for the calculation of different items based on the plinth area.
- The cost of the highway depends on the thickness of different layers, the number of lanes, the length of highways, and the facilities provided by the highways.
- The cost of acquiring the land also varies significantly with location for highway and railway projects.
- The approximate estimate for the irrigation projects requires a more detailed approach as it involves different hydrological, geological, and agricultural studies.

EXERCISE

Multiple Choice Questions

1. The approximate estimate is required to take the:

   (a) Administrative approval (b) Expenditure sanction

   (c) Technical sanction (d) none of the above

2. If the cost of construction of a school is 8000 ₹ per student. What will be the cost of the construction of a school with a capacity of 200 students?

   (a) 1400000 ₹ (b) 1600000 ₹

   (c) 1800000 ₹ (d) 2500000 ₹

   (Hint: Cost of construction = Cost of construction per student × Number of students)

3. If the carpet area is 1000 sq m. If the area of walls is 10 % of the carpet area and the circulation area is 30% of the carpet area, calculate the plinth area.

   (a) 300 sq m (b) 1100 sq m

   (c) 1300 sq m (d) 1400 sq m

   (Hint: $\boldsymbol{Plinth\ area = Carpet\ area + Area\ of\ walls + Circulation\ area}$)

4. If the plinth area is 100 sq m. Calculate the approximate number of bricks required for the construction.

(a) 21250 (b) 22576

(c) 23268 (d) 24457

(Hint: $Bricks(in\ 1000) = 0.226 \times Plinth\ area + 6.680$)

5. Which of the following cost is involved in the preparation of the approximate cost of a flexible pavement?

   (a) Cost of land acquisition (b) cost of layers

   (c) cost of drainage work (d) all of the above

6. The carpet area for the proposed construction is 1200 sq m and the area of walls is 20% of the plinth area. The circulation area is taken 25% of the plinth area. The approximate plinth area for the proposed construction will be:

   (a) 2112 sq m (b) 2152 sq m

   (c) 2182 sq m (d) 2252 sq m

   (Hint: See the solved numerical problem 2)

7. Circulation area represents the:

   (a) area of verandah (b) area of corridor

   (c) both of the above (d) none of the above

Answers of Multiple-Choice Questions

1. (a), 2. (a), 3. (d), 4. (c), 5. (d), 6. (c), 7. (c)

Short and Long Answer Type Questions

1. Explain the purpose of the approximate estimates. Also, describe the significance of the approximate estimates.
2. Describe the service unit method for the preparation of an approximate estimate. What are the different precautions taken during the preparation of the estimate by service unit method?
3. Write a short note on the following terms:

   a. Floor area

   b. Carpet area

   c. Plinth area

   d. Circulation area

4. Enlist the different precautions while preparing the approximate estimate using the cubical content method and plinth area method.
5. Explain the detailed procedure of the preparation of the approximate estimate for an irrigation project.
6. Explain the detailed procedure of the preparation of the approximate estimate for a water supply project.
7. Prepare the approximate estimate for a building with the following data:
    - Plinth area : 150 sq m
    - Plinth area rate : 35000 ₹/ sq m
    - Cost of water supply : 6 % of the cost of the building
    - Sanitary and electricity charges : 6.5 % of the cost of the building
    - Cost of decoration : 2.5 % of the cost of the building

    The contingencies are taken 8.5% and work charge establishment charges may be taken 10 %. Calculate the overall cost of the building project.

    (Hint: See the numerical problem 1)
8. Consider the following data for a building to be constructed :
    - Plinth area= 200 sq m
    - Height of the building from the ground floor= 3.5 m
    - Height of parapet wall= 0.8 m
    - Rate of per cu m volume of construction = 3100 ₹/ cu m Calculate the approximate cost of the building.

    (Hint: See the numerical problem 3)
9. Calculate the number of bricks and cement required for the construction with a plinth area of 1100 sq m. Use the empirical formula proposed by CBRI, Roorkee.
10. What is the typical bay method of approximate estimate? Explain the technique of typical bay in detail with relevant examples.

KNOW MORE

The calculation of the approximate cost of a structure requires an accurate and swift approach at the early stage of construction. However, in the approximate estimation of the bridges, the cost is calculated by multiplying the cost per running meter span by the span of the bridge which does not include the characteristics of the material used in the construction of the bridge. A new research published in the KSCE Journal of Civil Engineering proposes a rational approach for approximate estimation of the cost of bridges. To know more, scan the QR code.

REFERENCES AND SUGGESTED READINGS

1. Datta, B. N. *Estimating and Costing in Civil Engineering,* 1st Edition UBS Publisher Distributors Pvt. Ltd, New Delhi
2. Chakraborti, *M. Estimating, Costing, Specification & Valuation in Civil Engineering,* 1st Edition, Monojit Chakraborti, Kolkata
3. Peurifoy,R. L. & Oberlender, D. G. *Estimating Construction Costs*, 5th Edition, McGraw Hill Education, New Delhi.
4. Seeley, I.H. (1996). *Approximate Estimating*. In: Building Economics. Macmillan Building and Surveying Series. Palgrave, London. https://doi.org/10.1007/978-1-349-13757-2_6
5. Oh, C.D., Park, C. & Kim, K.J. (2013) *An approximate cost estimation model based on standard quantities of steel box girder bridge substructure*. KSCE Journal of Civil Engineering, 17, 877–885 (2013). https://doi.org/10.1007/s12205-013-0287-z

3

Detailed Estimates

UNIT SPECIFICS

Through this unit, we will discuss the following aspects:

- Detailed estimate and its purpose
- Data required for preparation of detailed estimate
- Unit quantity method and total quantity method of detailed estimation
- Long wall and short wall method and centre line method for finding out the quantities
- Bar bending schedule of different components
- Provisions in detailed estimate

To enhance the student's practical understanding, the long wall and short wall method and centre line method are discussed with solved numerical problems. Some of the important information is shared through individual plans of components of the building for a better understanding of the concepts.

This unit consists of exercises with long and short-term questions, multiple-choice questions, a list of references, and suggested reading to make the students learn through practice. Some extra information that may interest the students is provided in the form of QR codes. The students can scan these QR codes to reach the source of further knowledge on the different topics.

RATIONALE

This unit on the detailed estimates presents the different techniques for the calculation of the detailed probable cost of construction. The methods for finding out the quantities using the plans and sections of the building are discussed in detail. The long wall and short wall method and centre line method are discussed in detail with solved practical problems. The bar bending schedules of the different components are discussed with neat sketches. The content covered in this unit equips the students with procedures and methods of calculation of detailed costs.

The knowledge of preparation of detailed estimates is very important for civil engineering students and professionals. It helps the students understand and calculate the accurate probable cost of the construction projects using detailed drawings and plans. The concepts of long wall and short wall method and centre line method for taking out the quantities are essential to understand for a civil engineering student. The expertise in procedures of preparation of bar bending schedule is in demand in the civil engineering industry. The knowledge gained through this chapter will help the students in their civil engineering career and they will be able to calculate the detailed probable cost of a construction project with the detailed drawings and plans.

PRE-REQUISITES

The knowledge of different formats of estimation is required for understanding this chapter. An introductory understanding of reading the drawings and plans is required.

UNIT OUTCOMES

The list of outcomes of this unit is as follows:

U3-O1: Understand the requirements for the preparation of the detailed estimate

U3-O2: Calculate the quantities for the different components of the building

U3-O3: Prepare the bar bending schedule of the different building components

U3-O4: Understand the various provisions and rules of detailed estimation

Unit-3 Outcomes	EXPECTED MAPPING WITH COURSE OUTCOMES (1- Weak Correlation; 2- Medium correlation; 3- Strong Correlation)				
	CO-1	CO-2	CO-3	CO-4	CO-5
U3-O1	-	-	3	-	-
U3-O2	1	1	3	1	-
U3-O3	1	-	3	2	-
U3-O4	-	-	3	-	-

3.1 DETAILED ESTIMATE AND ITS PURPOSES

The approximate estimate provides a rough idea of the probable cost of the construction. However, there is a need to estimate an accurate probable cost that should be near the real construction cost. So, detailed estimates are prepared for determining the cost from the actual plans and drawings. A detailed estimate uses detailed designs and plans to determine the quantities of the different materials and their cost in the construction. The current rate of the materials is taken from the schedule provided by the public work departments. The probable cost of construction is calculated from these quantities and rates. A detailed estimate gives a clear picture of the probable cost of construction. The detailed estimates are prepared for the following purposes:

(1) The detailed estimates are prepared for the planning of the cost of the construction. It helps the owner to understand the probable costs involved in the construction of a project. The owner can prepare and modify the budget provisions based on the detailed estimate.

(2) An accurate detailed estimate may be used for securing the funds and loans for the construction project.

(3) The detailed estimates are used for scheduling the various activities of the project. It also helps in controlling the activities for the timely completion of the project.

(4) The detailed estimate is used for the efficient allocation of the resources at the different stages of the construction project.

(5) A detailed estimate helps in the preparation of the contract document. An accurate detailed estimate provides a clear and transparent breakdown of the cost.

(6) An accurate detailed estimate minimizes and solves the disputes among the stakeholders.

(7) A detailed estimate may provide a clear idea about the risks and uncertainty at the different stages of the projects in advance. These mitigation measures for these risks and uncertainty can be planned and taken in advance.

3.2 DATA REQUIRED FOR DETAILED ESTIMATE

The detailed estimates give an accurate prediction of the probable cost of construction. Detailed information about the design, material, and their cost is required for the preparation of the detailed estimate. The different type of data required for preparing the detailed estimate is given as follows:

(1) **Drawings and plans:** The detailed drawings and plans are required for the preparation of the detailed estimates. The drawings and plans are required for estimating the quantities of the materials.

(2) **Specifications:** The specifications are the standards that are used for maintain the quality of the work. The specifications are required to understand the quality and ratio of the different materials used for the construction. The general specifications and detailed specifications both are required. The specifications are discussed in Chapter 1 in detail.

(3) **Schedule of rates:** The rates of the items and different materials are required. The schedule of rates can be obtained from the Public Works Department (PWD) or any authorized agency at the state or central level.

(4) **Standing circulars:** The rate of some items of the work may not be available in the schedule. The standing circulars are the orders that give the rate of such items. These standard circulars are used for calculating the cost of items whose rates are not given in the schedule.

(5) **Goods and Services Tax (GST):** The knowledge of the taxes to be levied on the construction services is mandatory for the contractors and the estimator. Goods and Service Tax was introduced in 2014 and implemented in 2016 in India. It is a tax that is levied on services and goods. The different types of taxes like sales tax, service tax, excise duty, luxury tax, etc. are replaced by a single taxation system termed as 'GST'.

The GST in the construction services in India is 18% in most of cases. The GST on materials like tiles, refractory bricks, ceramic goods, glass material, etc. is also 18%. The GST on the different types of materials and services is given on the website of the Ministry of Finance. If more information is needed on the same, the QR code can be scanned.

(6) **Contingencies:** There are always some unforeseen events or tasks that may raise the cost of construction. It may be the unforeseen changes in the prices of the material or any other unpredictable event that may increase the cost of construction. The contingency is the fund/money taken as the percentage of the cost of construction that is kept to cover the unexpected costs. The contingencies are considered to cover the cost of incidental expenses that can't be classified in any head. In construction, the contingencies are kept from 3% to 5% of the construction cost.

(7) **Supervision charges:** The supervision of the construction work is an important task that ensures the construction is done according to the proposed design and specifications. The supervision services in construction include the cost of construction management, contract management, cost control, quality assurance, environment management, etc. The supervision charges are 4% to 10% of the project's cost.

(8) **Agency charges:** The agency charges are the internal cost of the construction company incurred due to the administrative and managerial operations. The rent of the office building, the salary of the office employees, the printing cost of the reports, etc. are the costs that come under the agency charges.

3.3 PROCEDURE FOR PREPARATION OF DETAILED ESTIMATE

The detailed estimate is prepared from the drawings, plans, and schedule of rate. The accuracy of the data used while preparing the detailed estimate ensures the calculation of the accurate probable cost of construction. So, the plans, drawings, mode of measurements, and schedule of rates should be accurate and updated at the time of preparation of the detailed estimate. The detailed estimate is prepared in the following steps:

(1) Taking out quantities

(2) Abstract of estimated cost

3.3.1 Taking Out Quantities

In the first step, the quantity of the different items of the work is calculated using the plans and drawings. The quantities are calculated from the dimensions of the different items. The quantities are calculated in the measurement form. The format of the measurement sheet is provided in Chapter 1 and is shown in Table 3.1.

Table 3.1: Standard format of measurement sheet

Item No.	*Description or Particulars of work*	*No.*	*Length*	*Breadth*	*Height or Depth*	*Content Quantity*	*Total Quantity*	*Remarks*
(1)	(2)	(3)	(4)	(5)	(6)	(7)	(8)	(9)

The content quantity is calculated by multiplying the dimensions (Column 4 to 6) and the total quantity is calculated by multiplying the content quantity (Column 7) by number (Column 3). The total quantity calculated in the measurement sheet is used for the calculation of the cost in the abstract sheet.

3.3.2 Abstract of Estimated Cost

The quantities calculated in the measurement sheet are now used for the calculation of the cost of construction of various items. The cost is calculated in the abstract form. The rate of items per unit is taken from the schedule of rates and the cost is calculated by multiplying these rates by the quantities of the items. A standard format of the abstract of cost is given in Table 3.2.

Table 3.2: Standard format of abstract sheet

Item No.	Description or particulars of the work	Quantity	Unit	Rate/unit item	Amount	Remarks
(1)	(2)	(3)	(4)	(5)	(6)	(7)

The quantity of the material or unit work is already calculated as discussed in the previous section and recorded in column 3. The unit of the quantity is mentioned in column 4. For example, the unit for plastering work is sq m and the unit for quantity of concrete is cu m. The rate of items of the work is the price. The standard unit rates can be obtained from the schedule of rates (SOR) published by the Public Works Department or any other competitive authority and mentioned in column 5. The amount is calculated by multiplying the quantity by the rate per unit item and mentioned in column 6. The cost for the contingencies, work charge establishment, sanitary and water supply charges, electrification charges, etc. are added to the calculated cost of construction to get the final cost of construction. These charges and the final cost of construction are shown in the face sheet. A standard format of the face sheet is given in section 1.8.3 of Chapter 1.

3.4 METHODS OF DETAILED ESTIMATE

The detailed estimate can be prepared using the unit quantity method or the total quantity method. Each method has its advantages and disadvantages. The methods for preparation of the detailed estimates are discussed as follows:

(1) **Unit quantity method:** The whole work is divided into the items of the work in the unit quantity method. The quantity of different items of work is calculated. The rate of the items of work is taken from the schedule of rates. The cost is calculated by multiplying the quantity by the cost per unit. For example, in the construction of a building, the earthwork is an item of work. The quantity of the earthwork is calculated from the drawings and plans. The cost is calculated by multiplying the quantity of the earthwork and its rate. The quantities and costs are calculated for all the items of the work and the final probable cost is calculated. Here, it is important to note that cost includes the cost of materials, labor, profit, etc. The main features of the unit quantity method are enlisted as follows:

- The unit cost method is simple to implement. The cost of the various items can be calculated using the quantities and rates.
- It is easy to detect and correct the errors in the unit quantity method.

- It is an accurate and transparent method. If the rates for the items are decided accurately, the probable cost can be calculated with good accuracy.
- The calculations in the unit quantity method become complex for large-scale projects.
- It is a time-consuming technique as there is a need for verification of the rates.
- The rates and quantities may change with time. So, the estimate should be revised from time to time for better accuracy.

(2) Total quantity method: In the total quantity method, the items of work are divided into five categories, and the cost is calculated for these categories. So, the total cost is calculated by adding the following costs:

- Cost of materials
- Cost of labor
- Cost of plant
- Overhead cost
- Profit of the contractor

The cost of each category is calculated by multiplying the quantity by its rate. The profit of the contractor is taken as some percentage of the cost. The cost of these categories is added to get the final cost of the construction. The total quantity method has the following features:

- The total quantity method provides a systematic approach to calculate the probable cost of construction.
- It is relatively difficult to detect the errors in the total quantity method as compared to the unit cost method.
- It is a systematic and quick technique. The complexity is reduced as the project is divided into five eategories only.

3.5 METHODS OF TAKING OUT QUANTITIES

At this stage, you are aware that the detailed estimate is prepared in two steps i.e. taking out quantities and abstracting. The calculation of quantities of different items of the work is very important as the cost of estimation depends on the accurate calculation of quantities. The quantities are calculated using the following methods:

(1) Long wall and short wall method

(2) Centre line method

3.5.1 Long Wall and Short Wall Method

The long wall and short wall method is also known as out to out and in to in method. It is sometimes termed as the 'general method' of detailed estimation. The work is divided into different items of work. Now, for each item of work, look at the plan and section drawings carefully. In the long wall and short wall method, walls of one direction are termed as long walls and the walls perpendicular to the long walls are known as short walls. Let's take an example of a room. The plan of the foundation of a single room is shown in Figure 3.1.

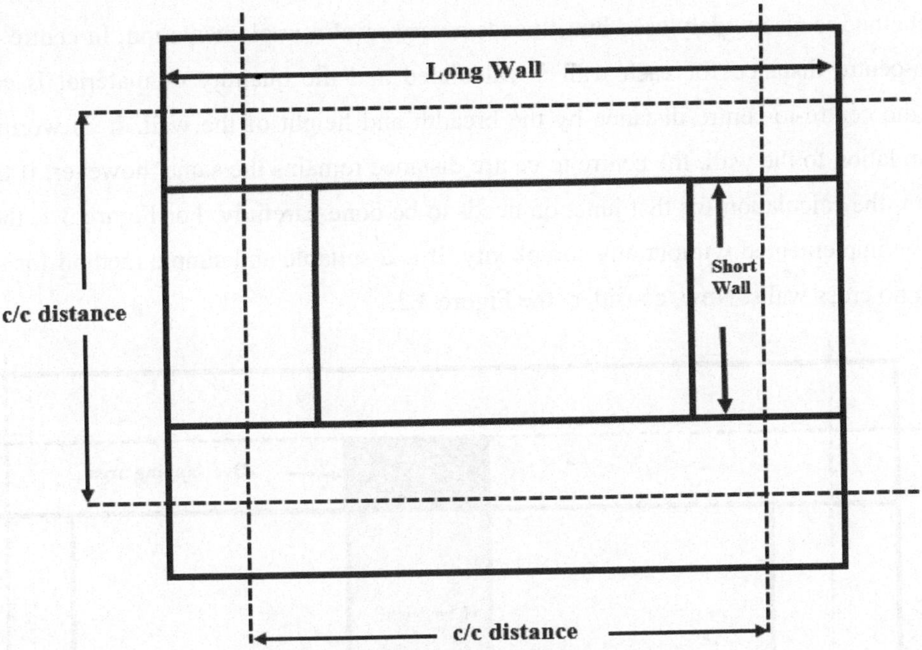

Figure 3.1: Long wall and short wall

There are four walls in the plan. Now the walls in one direction are termed as long walls and the walls in the other direction are called short walls. The center-to-center distance of each wall is calculated. From the figure, it can be seen that the length of the long wall is center-to-center distance plus the breadth of one short wall. Similarly, the length of the short wall is center-to- center distance minus the breadth of one long wall. The length of the long wall and short wall is given in the following equations:

$$Length\ of\ long\ wall = center\ to\ center\ distance + breadth\ of\ short\ wall$$

$$Length\ of\ short\ wall = center\ to\ center\ distance - breadth\ of\ long\ wall$$

Now breadth and height of the foundation are known. So, the quantity of the earthwork can be calculated by multiplying the length, breadth, and height. The same process can be done for determining the quantity of the short wall. Now from the section of the drawing, the dimensions of length, breadth, and height can be calculated for the other items of work like concrete and brickwork in the foundation, brickwork in walls, etc.

Important Note: It is important to note that the wall that is considered first is the long wall. However, its dimensions may be smaller than the short wall. So, it is just terminology and the students need not get confused. The concept will be more clear when you will see numerical problems related to these methods.

3.5.2 Centre Line Method

Centre line method is also widely used but it needs a very careful implementation. In centre line method, the centre-to-centre distance for each wall is calculated and the quantity of material is calculated by multiplying the centre-to-centre distance by the breadth and height of the wall. It is worth noting that from the foundation to the wall, the centre to centre distance remains the same, however, if there is some cross junction, the calculation for that junction needs to be done carefully. For Figure 3.1, the centre line method can be implemented without any complexity. It is a suitable and simple method for single rooms or plans with no cross walls. Now, consider the Figure 3.2.

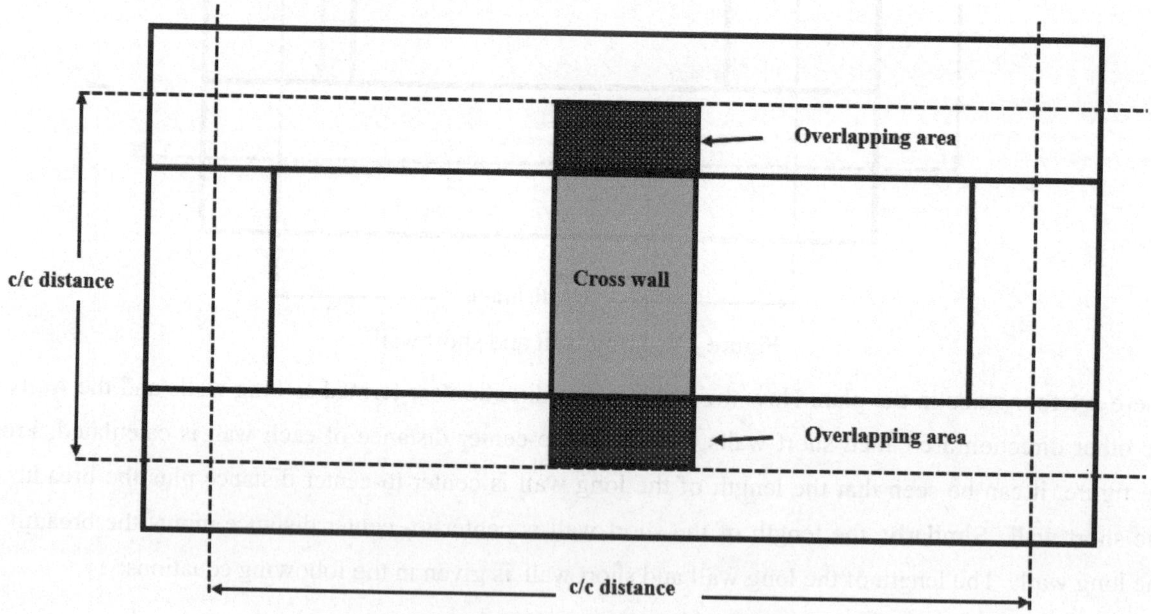

Figure 3.2: Structure with cross wall

In Figure 3.2, the calculation for the outer walls is simple. The centre-to-centre length for these walls is calculated and quantity is calculated by multiplying it by breath and height. The cross wall shown in the figure needs to be considered carefully as some areas are already considered while calculating quantities for the outer walls. The quantity will be overestimated if we calculate the quantity with centre-to-centre distance for the cross-wall. So, for the cross wall, the length is calculated by subtracting one breadth of the wall from the centre-to-centre distance. The concepts about the centre line method will also become clearer to the students when they understand the numerical problems related to it.

3.6 BAR BENDING SCHEDULE (BBS)

Reinforced Cement Concrete (RCC) consists of two items i.e. concrete and steel. There is a need for accurate calculation of the steel in the RCC sections. The quantity of the steel is calculated in the weight. In most cases, the quantity of the steel is estimated using detailed drawings and plans. The exact quantity of the steel is calculated including the hooks, cranks, etc. If detailed drawings or plans are not available, the quantity of the steel can be taken as a percentage of the concrete. Table 3.3 shows the calculation of the approximate quantity of steel as a percentage of concrete.

Table 3.3: Percentage of steel

S. No.	Construction Member	Quantity of Steel
1	Lintels and slabs	0.7% to 1%
2	Beams	1 % to 2%
3	Columns	1% to 5%
4	Foundation raft	0.5% to 0.8%

However, the calculations done through the percentage of steel are rough and the quantity of the steel estimated may not be exact. Some rules related to the bar bending schedule are explained as follows:

(1) For RCC works, the exposed surface of the reinforcement should be plastered, however, the quantity of the plastering is not calculated. In the case of Reinforced Brick (RB) works the plastering of the exposed surface is plastered with 12 mm thick plaster in 1:2 to 1:3 cement mortar. For RB works, the quantity of the plastering is measured and included in the detailed estimate.

Table 3.4: Rules for bar bending schedule

S. No.	Item	Dimension/formula
1	Bottom and top cover for slab	12 mm to 20 mm
2	Bottom and top cover for beam	20 mm to 25 mm
3	Side and end cover for steel bar	40 mm to 50 mm
4	Length of standard hook	$9.\Phi$
5	Length of a bar with hooks on both sides	$L + 2\times(9.\Phi) = L + 18.\Phi$

S. No.	Item	Dimension/formula
6	Length of 90° hook	6.Φ
7	Density of steel	7850 Kg/m^3
8	Weight of steel/meter	$Φ^2/162$ Kg/m

(2) There is a need to consider the reinforcement cover, hooks, bend-up, side cover, etc. carefully in the RCC construction. The rules for the cover, hook, and bend-up are summarized in Table 3.4. Here, L is the length of the bar and Φ is the diameter of the bar. For the calculation of weight, the diameter of the bar (Φ) is taken in mm and the weight comes in Kg/m.

(3) The quantity of binding wire should be estimated too. There is a need of 2.7 Kg binding wires of soft black iron for every 10 sq m of slab. For proper binding, 1 Kg binding wires are required for 1 quintal of steel.

A bar bending schedule provides information about the bars that are used in the construction. A BBS is presented in the tabular form. A good bar bending schedule should consist of the following information:

- Position of the bar in the structure or bar mark
- Shape of the bar
- Diameter and length of the bar
- Number of bars
- Total length of the bar
- Weight of each bar and total weight of the bars

3.6.1 Shapes of Bars

As already discussed, the estimation of steel in the construction should be accurate. The steel is measured in weight, so the dimensions of the steel bars should be accurately measured. The length of the hooked or bent bars depends on the hooks and the angle of bending. The bars can be shaped into different types discussed as follows:

(1) **Bar with hooks on both sides:** A bar with hooks on both sides is shown in Figure 3.3. The length of one hook is taken 9 times the bar diameter. The overall length of the bar is calculated by adding the length of hooks to the length of the straight bar.

$$\textit{Length of bar} = L + 18Φ$$

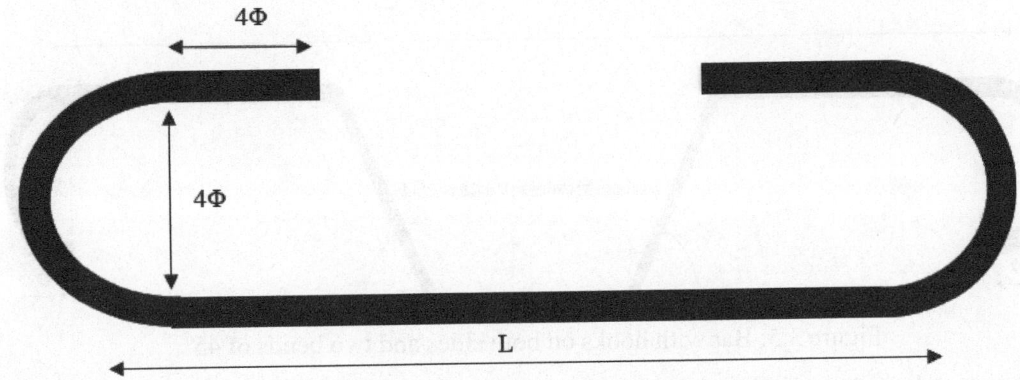

Figure 3.3: Bar with hooks on two sides

(2) **Bar with hooks on both sides and a bend of 45°:** A bar with hooks on both sides and with a bend of 45° at the central part is shown in Figure 3.4. The length of one hook is taken 9 times the bar diameter as discussed for the bar with hooks. The length of the bend is 0.42 D, where D is the height of the perpendicular formed in the bar as shown in the figure. So, the overall length is calculated by adding the length of the straight part, the hooks, and the length of the bends. Some authors use 0.45 D for the length of the bend, However, it is an approximate value for the length of the bend in the bar. The equation shows the overall length of a bar having a bend and hooks on both sides.

$$\textit{Length of bar} = L + 18\Phi + 0.42D$$

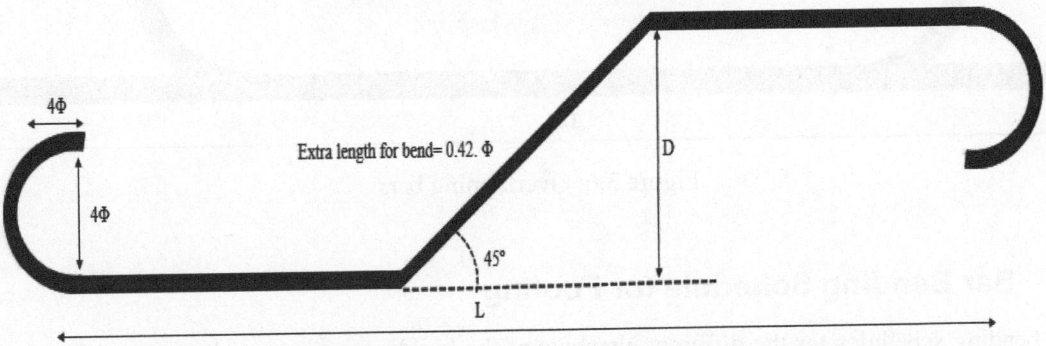

Figure 3.4: Bar with hooks on both sides and bend at the central part

(3) **Bar with two bends of 45° and hooks on both sides:** The calculation of length for the bar with two bends and two hooks is done similarly to the bar with two hooks and one bend, however, the length for the extra bend is added. Figure 3.5 shows the bar with two bends of height D and two hooks.

$$\textit{Length of bar} = L + 18\Phi + 0.84D$$

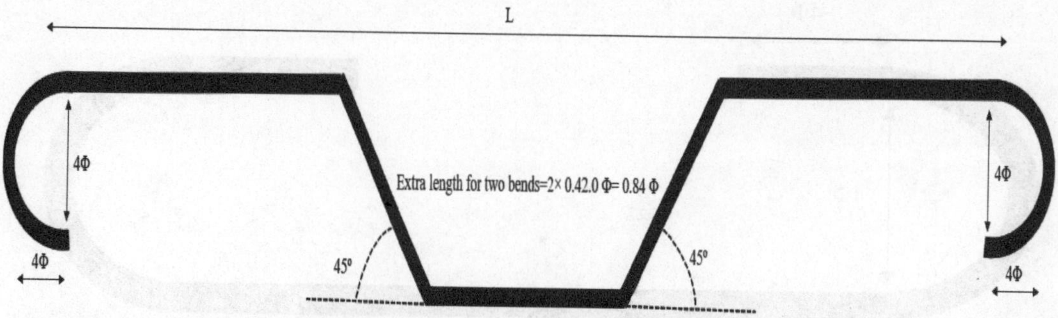

Figure 3.5: Bar with hooks on both sides and two bends of 45°

Overlapping of bars: Sometimes the sufficient length of the bar is not available. In this case, the bars are overlapped and the overlap length is calculated. The ends are hooked on both sides for tensile bars, however, in the case of compression bars there is no need to provide the hooks. The wiring is required carefully on both sides. Figure 3.6 shows the overlapping bars.The overlap length is given by the following equation.

$$\textit{Overlap length} = 40\Phi \textit{ to } 45\Phi$$

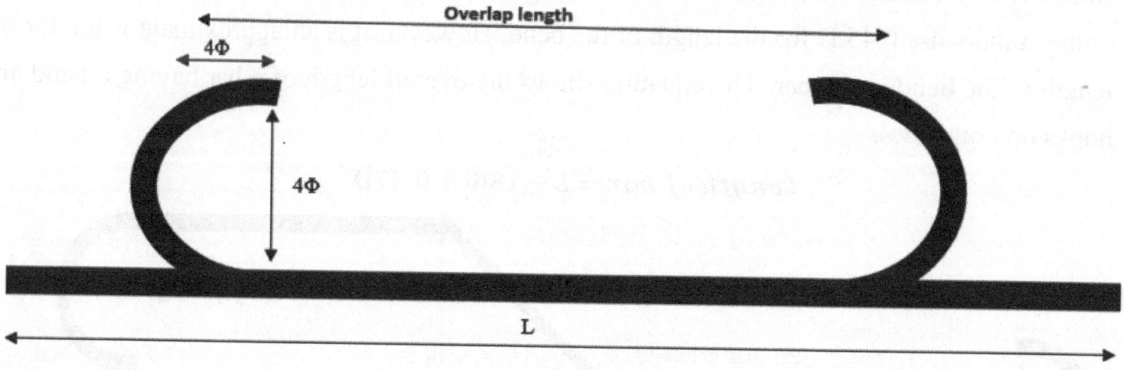

Figure 3.6: Overlapping bars

3.6.2 Bar Bending Schedule for Footing

The bar bending schedules for the different elements of the building are prepared using the drawings and plans. These can be understood by the examples. However, the basic concept of the calculation remains the same for most of the elements. Let's try to learn the preparation of a bar bending schedule for a footing through an example.

Solved Example 1: Let's prepare a bar bending schedule for an isolated footing shown in Figure 3.7.

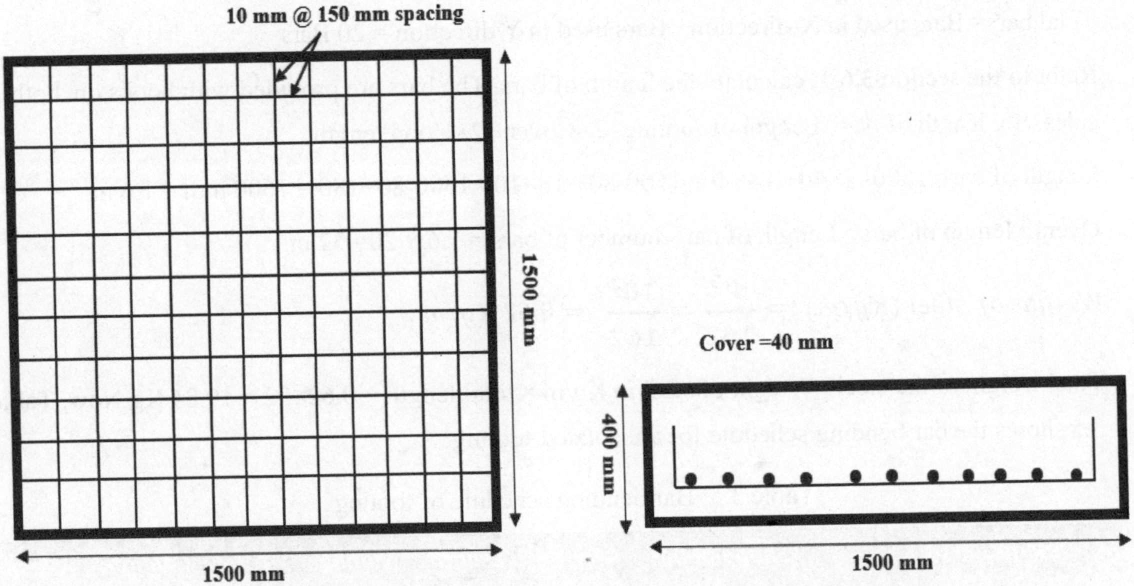

Figure 3.7: Drawing of footing showing reinforcement details

Solution:

Given data:

- Length of footing= 1500 mm
- Width of footing= 1500 mm
- Depth of footing= 400 mm
- Cover provided= 40 mm
- Diameter of bars in X- direction= 10 mm
- Diameter of bars in Y- direction= 10 mm
- Spacing in X-direction= 150 mm
- Spacing in Y-direction= 150 mm

Calculations:

- $Number\ of\ bars\ in\ X\ direction = \frac{Length\ of\ footing\ in\ Y\ direction - 2 \times Cover}{Spacing} + 1$

$$\frac{1500-2\times40}{150} + 1 = \frac{1420}{150} + 1 = 9.46+1 \approx 10 \text{ bars}$$

- **Due to the same dimensions of footing in the Y-direction and the same diameter of bars, the number of bars used in the Y-direction are same as X-direction i.e. 10 bars.**

- Total bars= Bars used in X-direction+ Bars used in Y direction = 20 Bars
- Refer to the section 3.6.1, calculate the length of bars. The bars are provided with hooks on both sides. So, length of bar= Length of footing- 2×Cover+ 2×Hoop length
- Length of bar= 1500- 2×40+ 18×Φ = 1500-80+ 18×10= 1500-80+180 = 1600 mm = 1.6 m
- Overall length of bars= Length of bar× number of bars = 1.6 × 20= 32 m
- $Weight\ of\ steel\ (Kg/m) = \frac{\Phi^2}{162} = \frac{10^2}{162} = 0.62\ Kg/m$
- Total weight of the steel= Weight of steel in Kg/m × total length = 0.62×32 = 19.84 Kg Now, Table 3.5 shows the bar bending schedule for an isolated footing:

Table 3.5: Bar bending schedule of footing

S. No.	Description of bar	Shape of bar	Length of bar (m)	Number of bars	Total length (m)	Weight (Kg/m)	Weight (Kg)
1	10 mm Φ bar in the X-direction and Y-direction of the footing		1.6	20	32	0.62	19.84

So, there is a need for 19.84 Kg of steel for this footing. In the bar bending schedule, the calculation is done for each category of bar. In the given example, the footing is square and has the same dimensions in the X and Y directions. Also, the bars used in both directions have the same diameter. That's why, the bar bending schedule can be presented in one row as shown in Table 5. However, if the dimensions of the footing and diameter of the bars are different in the X and Y directions, separate calculation is needed for each direction and total steel is calculated by adding the different weights calculated.

3.6.3 Bar Bending Schedule for Column

The columns may be of circular or rectangular type. There are vertical bars and steel stirrups used while reinforcing the columns. So, the bar bending schedule calculations are based on these vertical reinforcement and stirrups. The bars are overlapped and also the hooks are provided on both sides.

Generally, steel bars of 12 m in length are manufactured. Some of the basic concepts related to the bar bending schedule of the column that must be kept in mind are enlisted as follows:

(1) The length of the vertical bar is the sum of development length, height from the ground, floor heights, slab thickness, and overlap length.

(2) The lapping length for the bars is taken more than 40 to 50 times the diameter of the bars.

(3) The cutting length of the stirrups is taken as the sum of the perimeter of the stirrups, the length of bends, and the length of the hooks.

(4) There are two types of steel bars in a column i.e. vertical bars and stirrups. Both are calculated separately and presented in tabular form.

(5) A typical stirrup is shown in Figure 3.8. The number of bends and hooks can be seen in this figure.

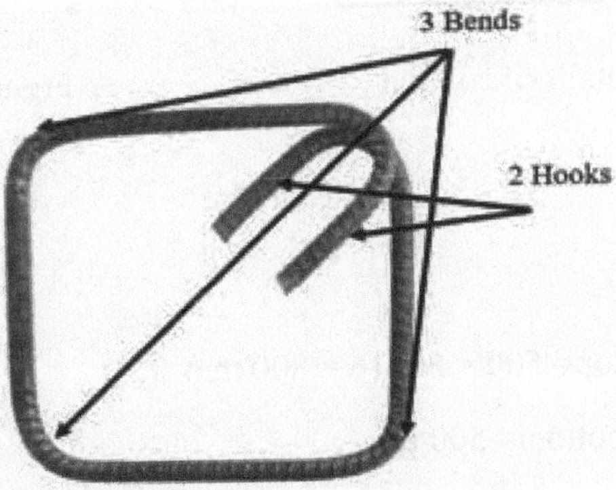

Figure 3.8: Bends and hooks

(6) For a hook, the length is taken 9 times the diameter of the bar.

(7) The length of bends at different angles is different. In Figure3. 9, the bends are 90º bends, For 90º bends, the length of the bend is taken 2 times the diameter of the bar. Table 3.6 shows the length of the bend at different angles.

Table 3.6: Bend length

S. No.	Angle of bend	Length of bend
1	45º	1×diameter of bar (Φ)
2	90º	2×diameter of bar (Φ)
3	135º	3×diameter of bar (Φ)
4	180º	4×diameter of bar (Φ)

Solved Example 2: Let's prepare a bar bending schedule for an RCC column shown in Figure 8:

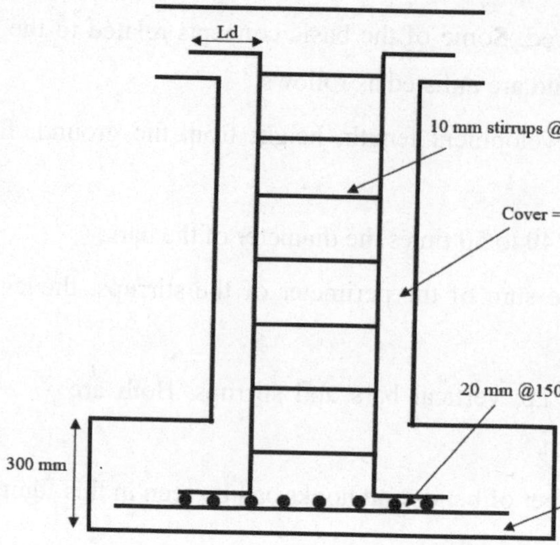

Figure 3.9 (a): Details of column

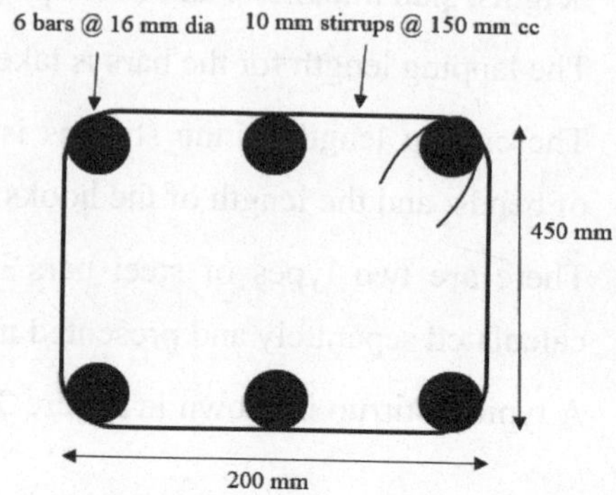

Figure 3.9(b) Section of stirrups

Solution:

Given data:

- Height= 4.5 m
- Development length at the top= 50Φ= 50×16 = 800 mm
- Development length at the bottom= 500 mm
- Vertical bar diameter: 16 mm
- Number of vertical bars: 6
- Diameter of stirrups: 10 mm
- Spacing of stirrups: 150 mm
- Cover in footing: 50 mm
- Cover in column: 40 mm
- Dimensions of stirrups: 450 mm× 200 mm

Calculations:

- Height from ground level= 4500 mm (Given)
- Development length= 50×Φ= 50×16= 800 mm (Should be more than 40 Φ)
- Length of vertical bar= Height+ development lengths+ vertical cover at the top of the column = - cover in footing-diameter of bars in footing **(See the drawing)**

= 4500+ 800+ 500+ 40- 50- 2×20 = 5750 mm= 5.75 m

- Total number of vertical bars= 6
- Total length of 6 vertical bars= 6×5.75= 34.5 m
- Weight per metre of bar = $\frac{\Phi^2}{162} = \frac{20^2}{162}$ = 2.47 Kg/m
- Total weight of vertical bars= Weight per metre × total length= 2.47×34.5= 85.215 Kg
- Number of stirrups required = $\frac{Length\ of\ vertical\ bar - development\ length}{spacing} + 1$

$$\frac{5750-800-500}{150} + 1 = 29.66+1 = 30.66 \approx 31$$

So, 31 stirrups are required. Now we will have to calculate the cutting length of the stirrups. See the b part of Figure 9.

- Cutting length of stirrups= Perimeter+ hook length-bend length
- Perimeter= 2×(200+450) = 2×650= 1300 mm

Now see Figure 8. You will be able to see the number of hooks and bends in for the stirrups. The stirrup has three bends and 2 hooks. So, the length of three 90º bends and 2 hooks is to be calculated.

- Length of bends= 3× (2×Φ) = 3×2×10= 60 mm
- Length of hooks= 2× (9×Φ)= 2×9×10= 180 mm
- Cutting length of stirrup= Perimeter+ hook length-bend length =1300+ 60+ 180

  = 1540 mm= 1.540 m
- Total stirrups required = 31 (already calculated)
- Total length of stirrups = cutting length× number of stirrups= 31×1.540= 47.74 m
- Weight per metre of bar = $\frac{\Phi^2}{162} = \frac{10^2}{162}$ = 0.617 Kg/m
- **Total weight of stirrups= 0.617×47.74= 29.45 Kg**

Table 3.7 shows the bar bending schedule for the column.

Table 3.7: Bar bending schedule of column

S. No.	Description of bar	Bar mark	Diameter of bar	Length of one bar (m)	Number of bars	Total length (m)	Unit Weight (Kg/m)	Weight (Kg)
1	20 mm Φ vertical bar	Vertical bars	20 mm	5.75	6	34.5	2.47	85.215
2	10 mm Φ stirrups @ 150 mm spacing	Tie bars/ stirrups	10 mm	1.54	31	47.74	0.617	29.45
Total weight of steel								**114.665**

So, **114.665 Kg** of steel is required for this column.

3.6.4 Bar Bending Schedule for Simple Beam and Lintel

The bar bending schedule of the lintel and simple beams is almost the same. The length of the main bars is calculated by deducting the covers and adding the hook length. The length of anchor bars is also calculated in the same way. The spacer bars may be provided in the perpendicular direction of the beam to provide support at the top. The length of these bars is calculated by deducting the side cover. The length and weight of stirrups in the beam are calculated in the same way as discussed in the column.

Solved Example 3: Let's prepare a bar bending schedule for a simple beam shown in Figure 3.10:

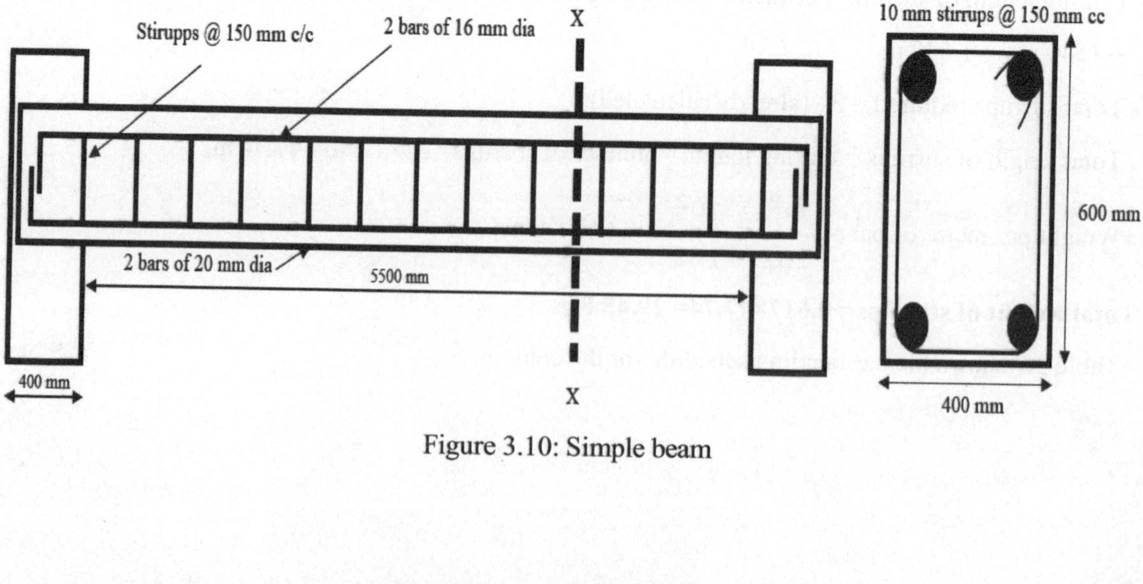

Figure 3.10: Simple beam

Solution:

Given data:

- Number of main bars at the bottom: 2
- Diameter of main bars at the bottom: 20 mm
- Number of main bars at the top: 2
- Diameter of main bars at the top: 16 mm
- Clear span of the beam= 5500 mm
- Clear cover= 30 mm (Not given. So, it is assumed)
- Diameter of bars in stirrups= 10 mm
- Spacing of stirrups= 150 mm
- Development length= 50Φ (Not given. So, the development length is assumed. It is normally taken more than 40 times the diameter)
- Breadth= 400 mm
- Width= 600 mm

Calculations:

- Cutting length of the bottom bar= clear span+ development length- clear cover

  = 5500+ 2×(50Φ)- 2×30 (Development length and clear cover of both sides is considered)

  = 5500+ 2×(50×20)- 60= 5500+ 2000- 60 = 7440 mm=7.44 m
- Number of bottom bars= 2
- Total length of the bars at bottom= 2×7.44 = 14.88 m
- Weight per metre of bar= $\frac{\Phi^2}{162} = \frac{20^2}{162}$ = 2.47 Kg/m
- **Total weight of bottom bars= Weight per metre × total length =2.47×14.88 = 36.75 Kg**
- Cutting Length of top bar= clear span+ development length- clear cover

  = 5500+ 2×(50Φ)- 2×30 = 5500+ 2×(50×16)- 60

  = 5500+ 1600-60 = 7040 mm = 7.04 m
- Number of bars at the top= 2
- Total length of the top bars= 2× 7.04 = 14.08 m

- Weight per metre of bar= $\frac{\Phi^2}{162} = \frac{16^2}{162}$ = 1.58 Kg/m
- **Total weight of bars at the top= 1.58×14.08 = 22.25 Kg**
- Number of stirrups required = $\frac{clear\ span\ of\ beam}{spacing} + 1 = = \frac{5500}{150} + 1$ = 37.67≈ 38 bars

So, 38 stirrups are required. Now we will have to calculate the cutting length of the stirrups. Calculate the cutting length of stirrups as discussed for the columns.

- Cutting length of stirrups= Perimeter+ hook length-bend length
- Perimeter= 2×(400+600) = 2×1000= 2000 mm

Table 3.8: Bar bending schedule of simple beam

S. No.	Description of bar	Bar mark	Diameter of bar	Length of one bar (m)	Number of bars	Total length (m)	Unit Weight (Kg/m)	Weight (Kg)
1	20 mm Φ at the bottom	Bars at bottom	20 mm	7.44	2	14.88	2.47	36.75
2	16 mm Φ at the top	Bars at top	16 mm	7.04	2	14.08	1.58	22.25
3	10 mm Φ stirrups @ 150 mm spacing	Stirrups	10 mm	2.24	38	85.12	0.617	52.52
Total weight of steel								**111.52**

Now see Figure 3.8 again to understand the configuration of the stirrups. The stirrup has three bends and 2 hooks. So, the length of three 90º bends and 2 hooks is to be calculated.

- Length of bends= 3× (2×Φ) = 3×2×10= 60 mm
- Length of hooks= 2× (9×Φ)= 2×9×10= 180 mm
- Cutting length of stirrup= Perimeter+ hook length-bend length =2000+ 60+ 180 = 2240 mm= 2.24 m
- Total stirrups required = 38 (already calculated)
- Total length of stirrups= cutting length× number of stirrups= 38×2.24= 85.12 m

- Weight per metre of bar = $\frac{\Phi^2}{162} = \frac{10^2}{162}$ = 0.617 Kg/m
- **Total weight of stirrups= 0.617×85.12= 52.52 Kg**

Table 3.8 shows the bar bending schedule for the column.

So, **111.52 Kg** of steel is required for the proposed simple beam.

The bar bending schedule for a lintel is also the same as that of a simple beam. The calculation of the cutting length and consideration for the bends and development length is given in the same way as done in the case of beams.

3.6.5 Bar Bending Schedule for Slabs

There are two kinds of slabs i.e. one-way slab and two-way slab. A one-way slab may be defined as a slab that has a length-to-breath ratio of 2 or more, while in two-way slabs the length-to-breadth ratio remains less than 2. The slab's length is its longer span and its breadth is its shorter span. The one-way slab is supported along its shorter span while the two-way slab is supported at both spans. Figure 3.11(a) shows the one-way slab while Figure 3.11 (b) shows the two-way slab.

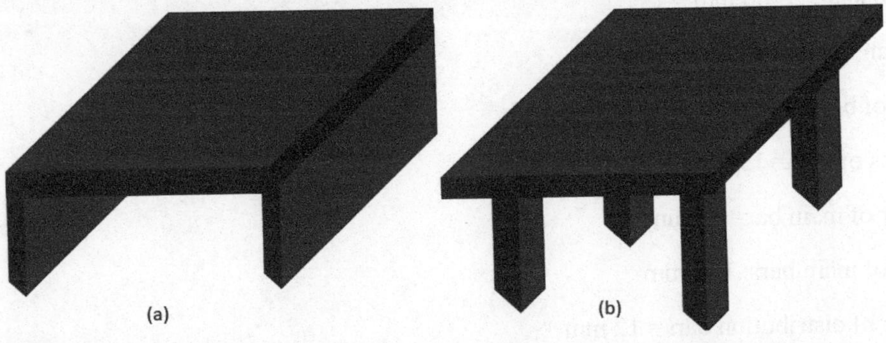

Figure 3.11: (a) One-way slab Figure 3.11: (b) Two-way slab

A one-way slab deflects only in one direction. So, the designing and analysis of such slabs is comparatively simple. However, a two-way slab is deflected in multiple directions, creating a complex design.

Solved Example 4: Let's prepare a bar bending schedule for a one-way slab shown in Figure 3.12:

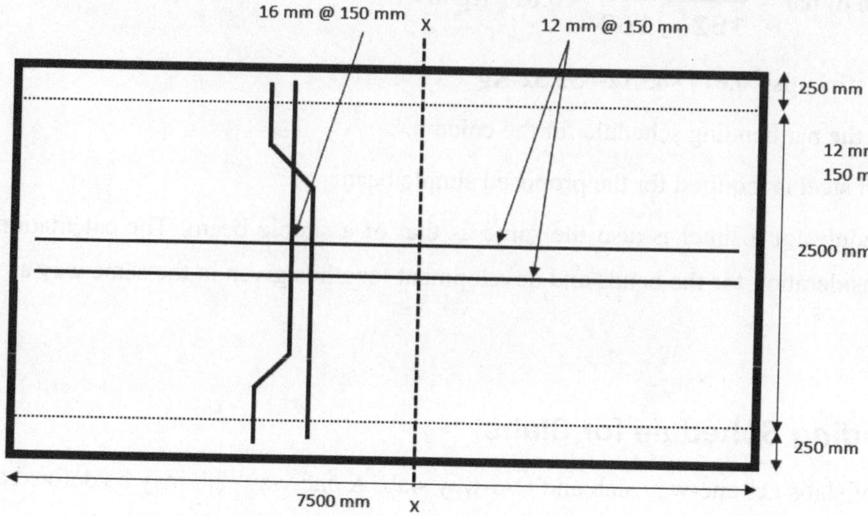

Figure 3.12: (a) Details of slab

Figure 3.12: (b) Section of slab

Solution:

Given data:

- Length of slab= 7500 mm
- Clear span of slab= 2500 mm
- Breadth of beam= 250 mm (on both sides)
- Thickness of slab= 150 mm
- Diameter of main bar= 16 mm
- Spacing of main bars: 150 mm
- Diameter of distribution bars= 12 mm
- Spacing of the distribution bars= 150 mm
- Clear cover (top and bottom) = 30 mm (Assume)
- Clear side cover= 35 mm (Assume)

Calculations:

Let's decide the type of the slab. Is it a one-way slab or a two-way slab?

- $\frac{\textit{Length of slab}}{\textit{Breadth of slab}} = \frac{7500}{2500} = 3$

The length-by-breadth ratio of the given slab is more than 2. So, it is a one-way slab.

- Number of main bars = $\frac{\textit{Length of slab}}{\textit{Spacing of main bars}} + 1 = \frac{7500}{150} + 1 = 50+1 = 51$ bars
- Cutting length of main bars= Clear span+ breadth of beam- clear cover+ inclined length-bend length

  Now for the main bar, calculate the different parameter.
- Clear span= 2500 mm (from figure)
- Breadth of beam= 250 mm on both sides (taken 2×250)
- Clear cover= 30 mm on both sides (taken 2×30)
- Inclined length= 0.42 D

  D can be seen in the figure given below:

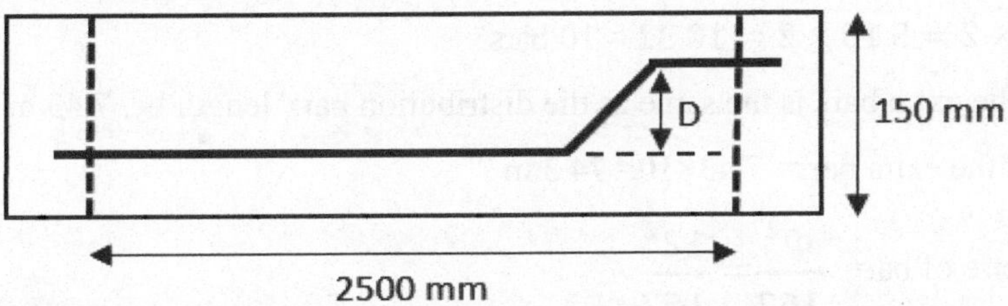

Figure 3.13: Elaborated figure for understanding the inclined length

- D= Breadth of slab- 2×clear cover- diameter of bar = 150-2×30-16 = 74 mm
- Bend length of 45° bend= 1×diameter of the bar (Note that the bar is bent at two points) Now put the values in the equation to calculate the cutting length of the main bars:
- Cutting length of main bars= Clear span+ breadth of beam- clear cover+ inclined length-bend length

  = 2500+ 2×250- 2×30+ 0.42×74- 2×16 (We are using 16 mm bar)

  = 2939.08 mm ≈ **2.939 m**
- Number of main bars= 51
- Total length of 51 bars= 51×2.939 = 149.889= 150 m
- Weight per metre of bar= $\frac{\Phi^2}{162} = \frac{16^2}{162} = 1.58$ Kg/m
- **Total weight of the main bars= 1.58×150= 237 Kg**
- Number of distribution bars= $\frac{\textit{Shoter span of slab}}{\textit{Spacing of main bars}} + 1 = \frac{2500}{150} + 1 = 16.667+1 = 17.667 \approx 18$ bars
- Cutting length of the distribution bar = long span - 2× clear side cover

  = 7500 - 2×35 = 7430 mm = 7.43 m

- Total length of the distribution bars= 7.43×18 = 133.74 m ≈ 134 m
- Weight per metre of bar = $\frac{\Phi^2}{162} = \frac{12^2}{162} = 0.89$ Kg/m
- **Total weight of the distribution bars= 0.89×134= 119.26 Kg**

  Extra bars are provided at the critical span of the slab. The length of the critical span on which the extra bars are provided is Shorter span/4.
- Span for providing extra bars= Shorter span/4= $\frac{2500}{4} = 625$ mm
- Number of extra bars required = $\left(\frac{\frac{Span}{4}}{spacing\ of\ bars} + 1\right) \times 2 = \left(\frac{625}{150} + 1\right) \times 2$

  $= (4.16 + 1) \times 2 = 5.16 \times 2 = 10.32 \approx 10$ bars

  The length of the extra bars is the same as the distribution bars' length i.e. 7.43 m
- Total length of the extra bars= 7.43×10= 74.3 m
- Weight per metre of bar= $\frac{\Phi^2}{162} = \frac{12^2}{162}$
- **Total weight of extra bars= 0.89×74.3 = 66.127 Kg**

Table 9 shows the bar bending schedule of a one-way slab:

Table 3.9: Bar bending schedule of slab

S. No.	Description of bar	Bar mark	Diameter of bar	Length of one bar (m)	Number of bars	Total length (m)	Unit Weight (Kg/m)	Weight (Kg)
1	16 mm Φ main bars	Main bars	16 mm	2.939	51	150	1.58	237
2	12 mm Φ distribution bars	Distribution bars	12 mm	7.43	18	134	0.89	119.26
3	12 mm Φ extra bars	Extr Bars	12 mm	7.43	10	74.3	0.89	66.127
			Total weight of steel					**422.387**

So, the total weight of steel bars required for the given slab is **422.387 Kg.**

3.7 PROVISIONS IN DETAILED ESTIMATE

The calculation of the construction cost in a detailed estimate is discussed in detail in the previous sections. Some other types of costs are calculated as a percentage of the cost of construction. The different provisions in detailed estimates should be understood clearly for calculating the final probable cost. The lump sum funds for the different purposes are kept in the budget allocation. The different provisions in a detailed estimate are enlisted as follows:

(1) Contingencies

(2) Work-charged establishment

(3) Percentage charges

(4) Water supply and sanitary charges

(5) Electrification charges

(1) Contingencies: The contingencies may be defined as the funds that are kept to cover the cost of uncertain events or circumstances that can occur during project implementation. The cost of contingencies is taken as a percentage of the cost of construction. The cost of contingencies included in the detailed estimate helps in the management of risk and uncertainty of the construction project. The major provisions for the contingencies and their costs are enlisted as follows:

- The contingencies are taken as a percentage of the cost of construction.
- The cost of contingencies in construction is usually taken from 3% to 5% of the cost of construction.
- The contingency funds can't be used for variation in design or repair work. These funds are used for unforeseen events or circumstances that may occur during the construction of the project.

(2) **Work-charged establishment:** There are costs of execution of the project that include the salaries of the temporary employees, supervisors, staff, etc. These employees are paid based on the work estimate on which they are assigned. The provisions for work-charged establishment are enlisted as follows:

- The work-charged staff are the temporary employees that may be hired to perform the particular tasks.
- The cost of work-charged establishment is taken from 1.5% to 2% according to the Public Works Department (PWD).
- The work-charged staff can be terminated by giving one month's notice in most of the cases.

(3) **Percentage charges:** The probable cost of the construction is calculated from a detailed estimate. The different charges that are taken as a fraction of the probable cost are known a the percentage charges. For example, the cost of contingencies and work-charged establishment are the percentage charges that are already discussed. Some more percentage charges are given as follows:

- For the cost of tools and machinery, the provision of 1% to 1.5% of the estimated cost of construction is considered.
- The contractor's profit is added at the rate of 10% of the estimated cost of construction.
- The profit of the department varies from 10% to 15% of the estimated cost.

(4) **Water supply and sanitary charges:** The water supply and sanitary charges are also calculated as the percentage of the total cost of construction. The lump sum cost of the water supply and sanitary services is considered in the detailed estimate. The provision of keeping 8% of the estimated cost of construction for the water supply and sanitary charges.

(5) **Electrification charges:** The installation of electrical facilities in the building and cost of electricity is also included separately like water supply and sanitary charges. The electrification charges are taken around 8% of the estimated cost.

3.8 PRIME COST

The quantities of the items can be calculated in the detailed estimate, however, every detail or item may not be calculated. The details of such items are not worked out at the time of the tendering process and are included in the building in the future. For example, the fitting of door handles, water supply fittings, etc. is decided at the time of actual fitting. The actual cost of such items at the shop is known as prime cost. The cost of installation or carrying is not included in the prime cost. The contractor provides such items as per instructions of the Engineer-in-Chief. The important features of the prime cost are enlisted as follows:

(1) The prime cost refers to the actual cost of the item paid by the contractor to the merchant. The cost of carrying or installation is not included.

(2) The actual cost of the items is paid to the contractor. However, the carriage charges may be provided separately.

(3) The contractor can't take any profit on the articles of prime cost.

(4) The prime cost should be realistic and must be checked with the current prices of the items.

(5) The instructions about the quality and prices of such items are given by the Engineer-in- Chief.

3.9 PROVISIONAL SUM

There may be some special type of work whose exact cost is not available at the time of preparing the estimate. The cost of the specialized works like installation of refrigeration facility, air conditioning facilities, etc. may not be known at the time of preparation of the estimate. The allowance provided for such specialized works is known as the provisional sum. The provisional sum includes the cost of items that can't be taken off during estimation or that are to be measured at the site.

3.10 PROVISIONAL QUANTITIES

If additional quantities of some items are required during the construction due to the change in site condition or any other reason, then these quantities are considered separately and kept under the heading of provisional quantities. The earthwork is a major example of the provisional quantities. The quantity of earthwork varies usually due to the site conditions.

3.11 SPOT ITEMS/SITE ITEMS

The quantities of some items can't be calculated without studying or visiting them. Such items that can't be estimated without visiting or studying are known as the spot items or site items. The estimation of cost for such items should be done only after the inspection of these items. Some examples of the spot items are given as follows:

(1) The cost of demolition of the structures can only be predicted after visiting the sites and studying the site conditions.

(2) The cost of site clearance can only be predicted by visiting the site and studying the condition of the site.

(3) Construction of openings in the existing walls also falls in the category of the spot items. A precise detail of the spot items and their description should be provided to the contractor.

3.12 BILL OF QUANTITIES

The bill of quantities is a document comprising the details of the items, materials, labor, etc. required to complete the construction project. The bill of quantities provides the quantities measured from the drawings which can be used for calculation of the cost of construction. The bill of quantities is prepared for the following reasons:

(1) The bill of quantities helps in efficient estimation. The details given in the bill of quantities are used for the preparation of the accurate budget.

(2) The bill of quantities helps in proper planning and scheduling of the construction project. It gives the idea of the different items that are needed to complete the different stages of the project.

(3) The bill of quantities brings transparency among the stakeholders of the project. It also reduces the possibility of disputes between the client and the contractor.

(4) As the idea of the probable cost at different stages is known through the bill of quantities, it helps in controlling the cost too.

(5) The bill of quantities helps the contractors to propose a realistic bid.

3.12.1 Rules of Preparation of Bill of Quantities

The ultimate goal of the preparation of the bill of quantities is to document the cost of the different items of the work based on the measurement and specifications. The pricing of the items is done based on the cost of material, labor, equipment, and profit of the contractor. The following rules are followed for pricing the different items:

(1) The cost of material is calculated through the quotations invited from the different suppliers. The specifications and proposed prices by the suppliers are analyzed to finalize the cost of the material. However, in India, the schedule of rates is used for the calculation of the cost of materials.

(2) The price of equipment and machinery to be used at the different stages of the construction is taken as a percentage of the cost of material. However, the price of equipment depends on whether the machinery and equipment are rented or owned.

(3) The cost of skilled and unskilled labor is included in the bill of quantities. The payment to the labor is done based on the schedule of rates.

(4) As the cost of material, machinery, labor, etc. is decided, the profit of the contractor can be added as a percentage of the total cost. A typical format of the bill of quantities is shown in Table 3.10.

Table 3.10: Bill of quantities

Provisional Sum						
Item No.	***Description or particulars of the work***	***Unit***	***Quantity***	***Rate***	***Amount***	***Remarks***
Construction Work						
Item No.	***Description or particulars of the work***	***Unit***	***Quantity***	***Rate***	***Amount***	***Remarks***

(5) The first part is used to calculate the provisional sum. The items of the work in which the contractor can't take any profit are placed under the heading of the provisional sum. The cost of insurance premiums, lab testing, etc. is calculated in the provisional sum. The details of the quantities of the construction work through the measurement of the drawings and their cost are placed under the construction work heading. A typical bill of quantities can be seen by scanning the QR code.

3.13 CONCLUSIONS FROM THE CHAPTER

An understanding of the detailed estimate and its procedures is essential for civil engineering students and professional quantity surveyors. In this chapter, the student understands the plans and drawings required for the preparation of the detailed estimates. The different techniques of calculation of quantities are discussed in a detailed manner with practical numerical problems. The bar bending schedule of the different components of building is also presented with practical problems. A brief discussion of the provisions of the detailed estimation is also included at the end of the chapter.

After the completion of this chapter, the students will be able to calculate the quantities of the materials in the buildings. They will be able to prepare the bar bending schedules for the sites. The students will acquire sufficient knowledge about the provisions of the detailed estimates. The students will be ready to understand the estimates of the other civil engineering works at this stage. It is important to learn about the earthwork calculation, estimation of septic tanks, etc. In the current market situation, the role of software in estimation and costing has increased significantly. The next chapter deals with all these concepts.

SOLVED NUMERICAL EXAMPLES

Problem 1:

Prepare an estimate for a one-roomed building using long wall and short wall method with the following quantities:

- Earthwork of foundation
- Concrete in foundation
- Brickwork in foundation and superstructure

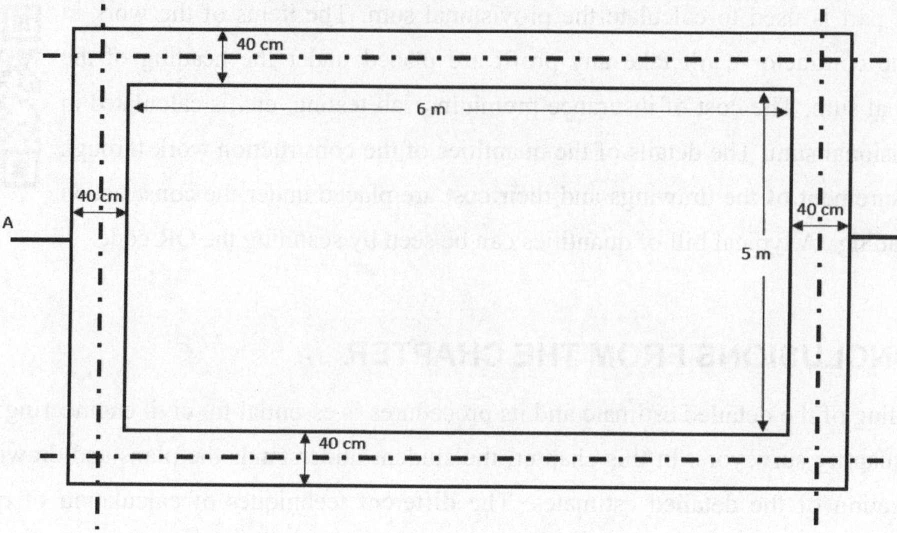

Figure 3.14 (a) Plan of the building

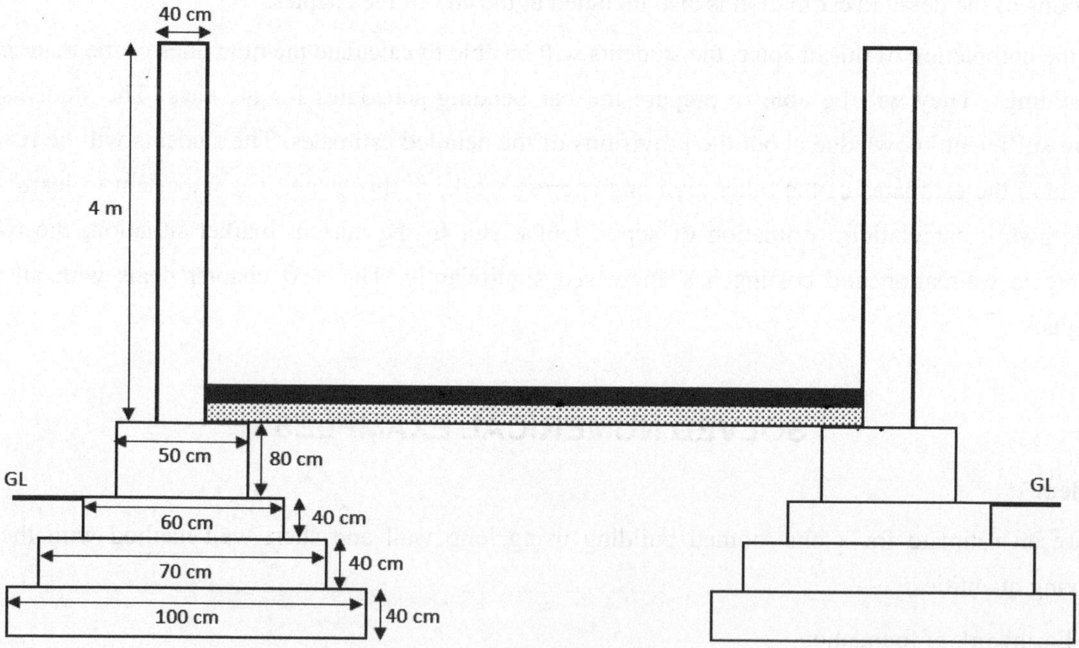

Figure 3.14 (b): Section A-A

Solution:

Given data:

- Internal length of the long wall: 6 m
- Internal length of the short wall: 5 m

- Width of the wall: 40 cm: 0.4 m
- Depth of lime concrete in base: 40 cm
- Height of first footing and second footing: 40 cm
- Height of the plinth: 80 cm
- Height of ground level: 40+40+40 = 120 cm (from section A-A)

Calculation:

- Centre to centre length of the long wall (L_x) = $6 + \frac{1}{2} \times 0.4 + \frac{1}{2} \times 0.4 = 6.4$ m (See the plan)
- Centre to centre length of the short wall (L_y) = $5 + \frac{1}{2} \times 0.4 + \frac{1}{2} \times 0.4 = 5.4$ m (See the plan)
- Now remember the following rule:
  - ➢ Add the width of one wall from the centre-to-centre distance of the long wall.
  - ➢ Subtract the width of one wall from the centre-to-centre distance of the short wall.
- Calculation for the excavation:
  - ➢ Understand the calculations for the earthwork carefully. From the section A-A, the breadth of the foundation is 100 cm or 1 m. The excavation of the earth is done up to ground level. The excavation depth is up to ground level, i.e. 120 cm or 1.2 m.
  - ➢ The plan for the foundation will be seen as shown in the Figure.

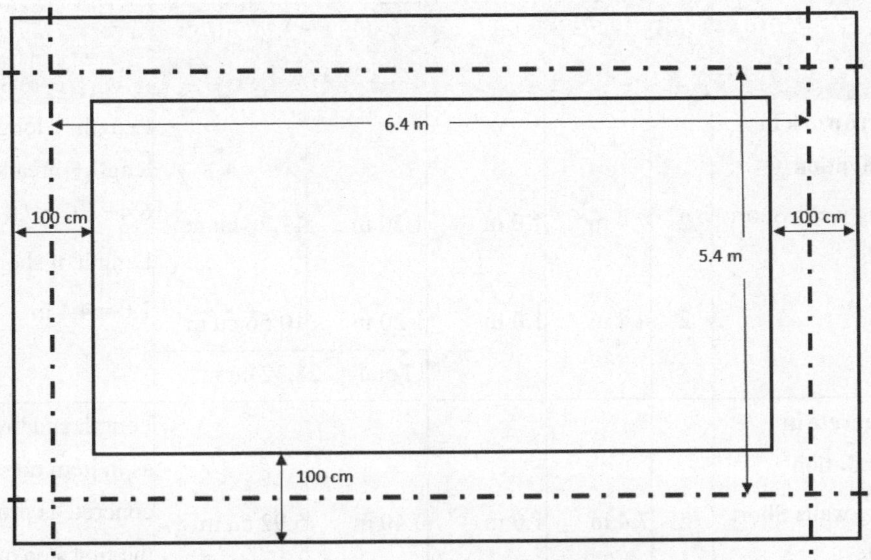

Figure 3.14 (c): Plan for foundation

- So, the length of the long wall= c/c length+ breadth of the wall= 6.4+ 1.0 = 7.4 m
- So, the length of the short wall= c/c length- breadth of the wall= 5.4- 1.0 = 4.4 m

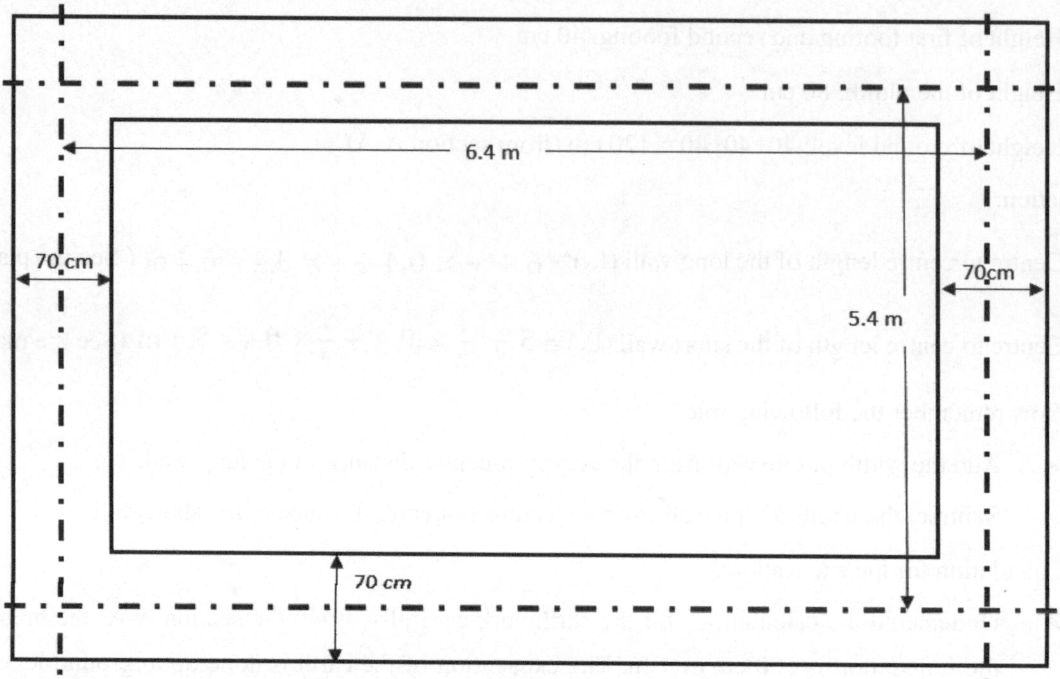

Figure 3.14 (d) Plan for first footing

Table 3.11: Estimation for one-roomed building

Item no.	Description of item	No	Length	Breadth	Height	Quantity	Explanation
1	**Earthwork in excavation** Long walls Short walls	2 2	7.4 m 4.4 m	1.0 m 1.0 m	1.20 m 1.20 m	17.76 cu m 10.56 cu m	Length long wall= c/c length+ breadth of the wall= 6.4+ 1.0 = 7.4 m Length of short wall= 5.4- 1.0 = 4.4 m
					Total =	28.32 cu m	
2	**Concrete in foundation** Long walls Short walls	2	7.4 m	1.0 m	0.40 m	5.92 cu m	Length and breadth are same as in item no. 1 but the concrete is provided only on the first step of 40 cm depth.

		2	4.4 m	1.0 m	0.40 m	3.52 cu m	
					Total =	9.44 cu m	
3	**Brickwork in foundation** **Long walls** First Footing Second footing Plinth walls **Short walls** First Footing Second footing Plinth walls						Length of long wall for first footing= 6.4+ 0.70 = 7.1 m Length of short wall for first footing = 5.4- 0.7 = 4.7 m
		2	7.1 m	0.70 m	0.40 m	3.976 cu m	
		2	7.0 m	0.60 m	0.40 m	3.360 cu m	Length of long wall for second footing= 6.4+ 0.60 = 7.0 m
		2	6.9 m	0.50 m	0.80 m	5.520 cu m	Length of short wall for second footing = 5.4- 0.6 = 4.8 m
		2	4.7 m	0.70 m	0.40 m	2.632 cu m	
		2	4.8 m	0.60 m	0.40 m	2.304 cu m	Length of long wall for plinth= 6.4+ 0.50 = 6.90 m
		2	4.9 m	0.50 m	0.80 m	3.920 cu m	Length of short wall for plinth = 5.4- 0.5 = 4.9 m
					Total =	27.712 cu m	
4	**Brickwork in superstructure**						Length of long wall for spuperstructure= 6.4+ 0.40 = 6.80 m
	Long walls Short walls	2	6.80 m	0.40 m	4.0 m	21.76 cu m	Length of short wall for first footing = 5.4- 0.40= 5.0
		2	5.00 m	0.40 m	4.0 m	16.00 cu m	
					Total =	37.76 cu m	

- Understand the calculations for the concrete in the foundation carefully. The length of the long wall and short wall remains the same as that for the earthwork, however, the concrete is provided only for 40 cm in height. After that, brickwork is provided. So, only the depth for the concrete work is changed in the case of the calculation of concrete in the foundation. For the first footing, the plan is redrawn according to the dimensions of the footing. For the first footing, the plan will be as shown in the Figure above.
- Length of long wall for first footing= 6.4+ 0.70 = 7.1 m
- Length of short wall for first footing = 5.4- 0.70 = 4.7 m
- All the other calculations including the calculation for the second footing, for the plinth and superstructure are performed similarly. The centre-to-centre length remains the same and the plan changes at the different stages.
- The detailed calculations are shown in the Table.

Problem 2:

Prepare an estimate for a one-roomed building using centre line method with the following quantities:

- Earthwork of foundation
- Concrete in foundation
- Brickwork in foundation and superstructure

The plan and section of the building are shown in the figure. The base of 40 cm is provided with lime concrete.

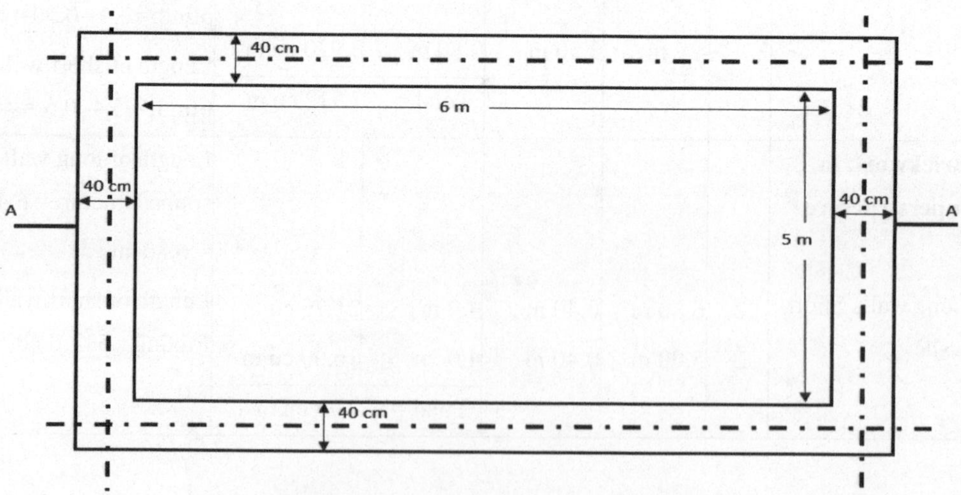

Figure 3.15 (a): Plan of the building

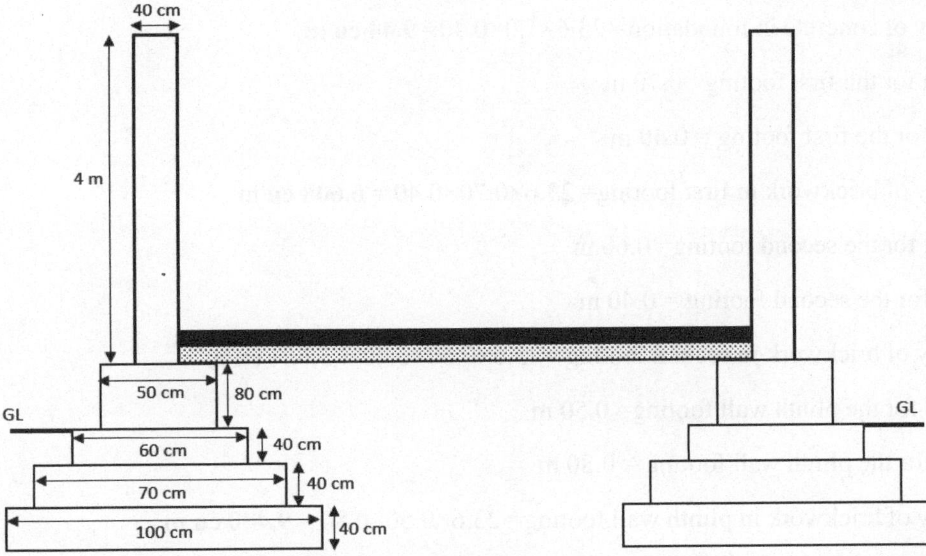

Figure 3.15 (b): Section A-A

Given data:

- Internal length of the long wall: 6 m
- Internal length of the short wall: 5 m
- Width of the wall: 40 cm: 0.4 m
- Depth of lime concrete in base: 40 cm
- Height of first footing and second footing: 40 cm
- Height of the plinth: 80 cm
- Height of ground level: 40+40+40 = 120 cm (from section A-A)

Solution:

- Centre to centre length of the long wall= 6+ 0.40 = 6.40 m
- Centre to centre length of the short wall= 5+ 0.40 = 5.40 m
- Total centre to centre length of the building= 2× (6.40+ 5.40)= 23.6 m
- Breadth for the earthwork= 1.0 m
- Height for the earthwork= 1.2 m
- Quantity of earthwork= Length×Breadth×Height= 23.6×1.0×1.2 = **28.32 cu m**
- The breadth of the concrete in foundation= 1.0 m
- The height of the concrete in foundation= 0.40 m

- Quantity of concrete in foundation= 23.6×1.0×0.40= **9.44 cu m**
- Breadth for the first footing= 0.70 m
- Height for the first footing = 0.40 m
- Quantity of brickwork in first footing= 23.6×0.70×0.40 = **6.608 cu m**
- Breadth for the second footing= 0.60 m
- Height for the second footing = 0.40 m
- Quantity of brickwork in second footing= 23.6×0.60×0.40 = **5.664 cu m**
- Breadth for the plinth wall footing= 0.50 m
- Height for the plinth wall footing = 0.80 m
- Quantity of brickwork in plinth wall footing= 23.6×0.50×0.80 = **9.440 cu m**
- Total quantity of brickwork in the footing= Brickwork in first footing+ Brickwork in second footing+ Brickwork in footing for plinth wall

  = 6.608+ 5.664+ 9.440 = **21.712 cu m**
- Breadth for the wall of superstructure= 0.40 m
- Height for the wall of superstructure = 4.0 m
- Quantity of brickwork in superstructure= 23.6×0.40×4.0 = **37.76 cu m**

Important Note: Compare the results calculated by the centre line method and long wall and short wall method. We have solved the same numerical with these two methods. The final quantities calculated with both methods are the same. The centre-line method should be implemented very carefully whenever there is a cross wall.

Problem 3:

The plan and section of a two-roomed building are shown in the figure. The base of 40 cm is provided with lime concrete.

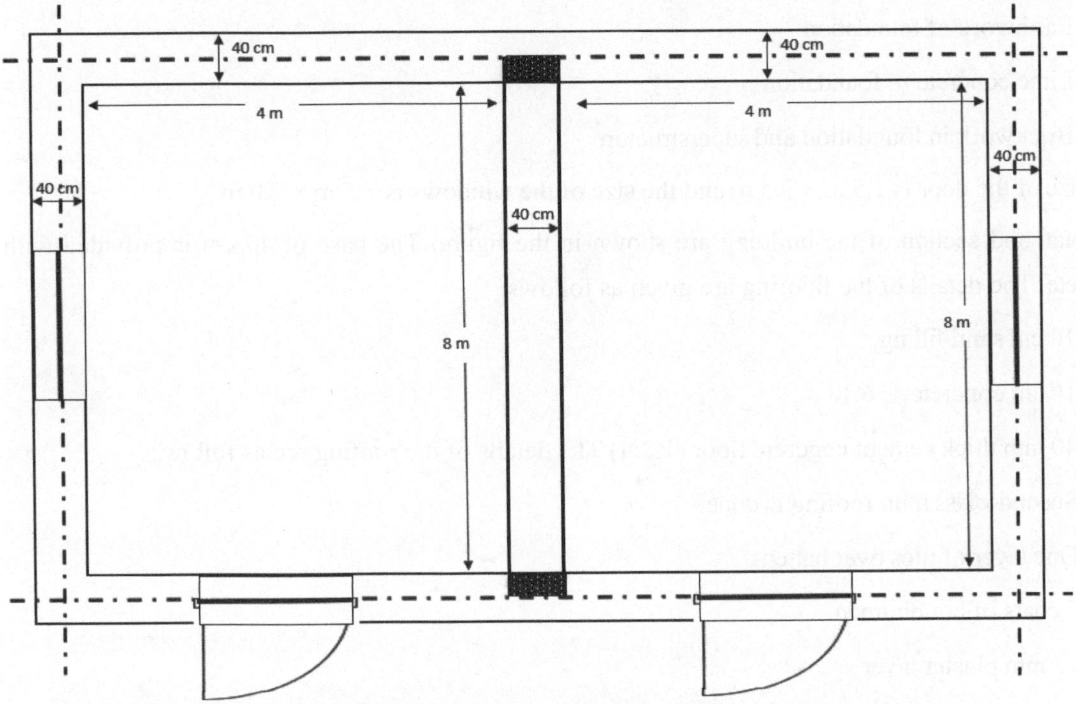

Figure 3.16 (a): Plan of the building

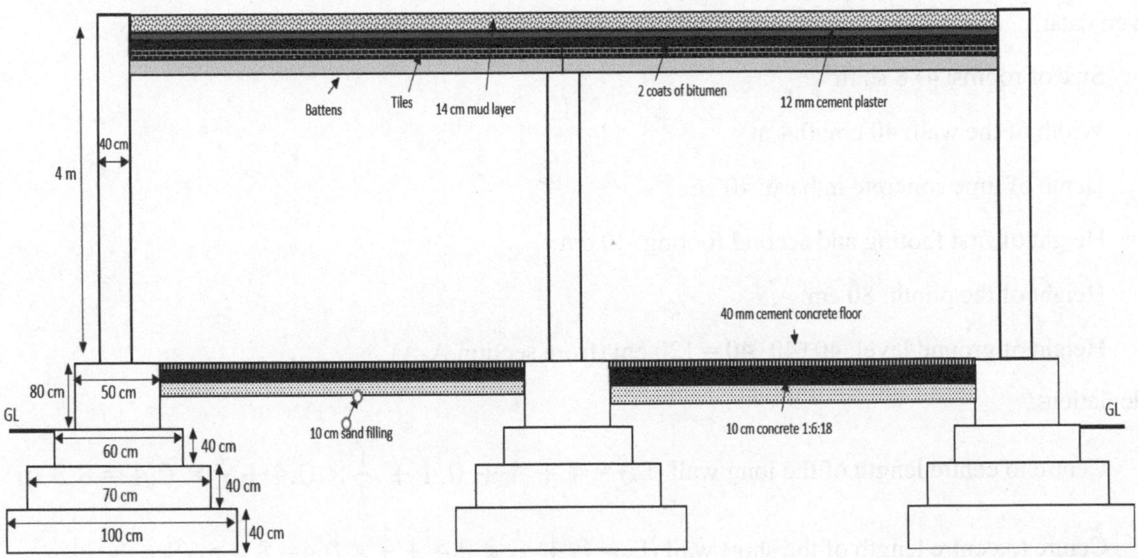

Figure 3.16 (b): Section of the building

Prepare an estimate for the shown two-roomed building using the long wall and short wall method with the following quantities:

- Earthwork of foundation
- Lime concrete in foundation
- Brickwork in foundation and superstructure

The size of the door is 1.5 m × 2.5 m and the size of the windows is 1.5 m × 2.0 m.

The plan and section of the building are shown in the figure. The base of 40 cm is provided with lime concrete. The details of the flooring are given as follows:

- 10 cm sand filling
- 10 cm concrete 1: 6:18
- 40 mm thick cement concrete floor (1:2:4) The details of the roofing are as follows:
- Second-class mud roofing is done
- One layer of tiles over battens
- 2 coats of hot bitumen
- 12 mm plaster layer
- 14 cm mud layer

Solution:

Given data:

- Size of rooms: 4×8 sq m
- Width of the wall: 40 cm: 0.4 m
- Depth of lime concrete in base: 40 cm
- Height of first footing and second footing: 40 cm
- Height of the plinth: 80 cm
- Height of ground level: 40+40+40 = 120 cm (from section A-A)

Calculations:

- Centre to centre length of the long wall (L_x) = $4 + 4 + 0.4 + \frac{1}{2} \times 0.4 + \frac{1}{2} \times 0.4 = 8.8$ m
- Centre to centre length of the short wall (L_y)= $8 + \frac{1}{2} \times 0.4 + \frac{1}{2} \times 0.4 = 8.4$ m (See the plan)

The only major point to understand is that there are two long and three short walls. The plan for the earthwork will look as follows:

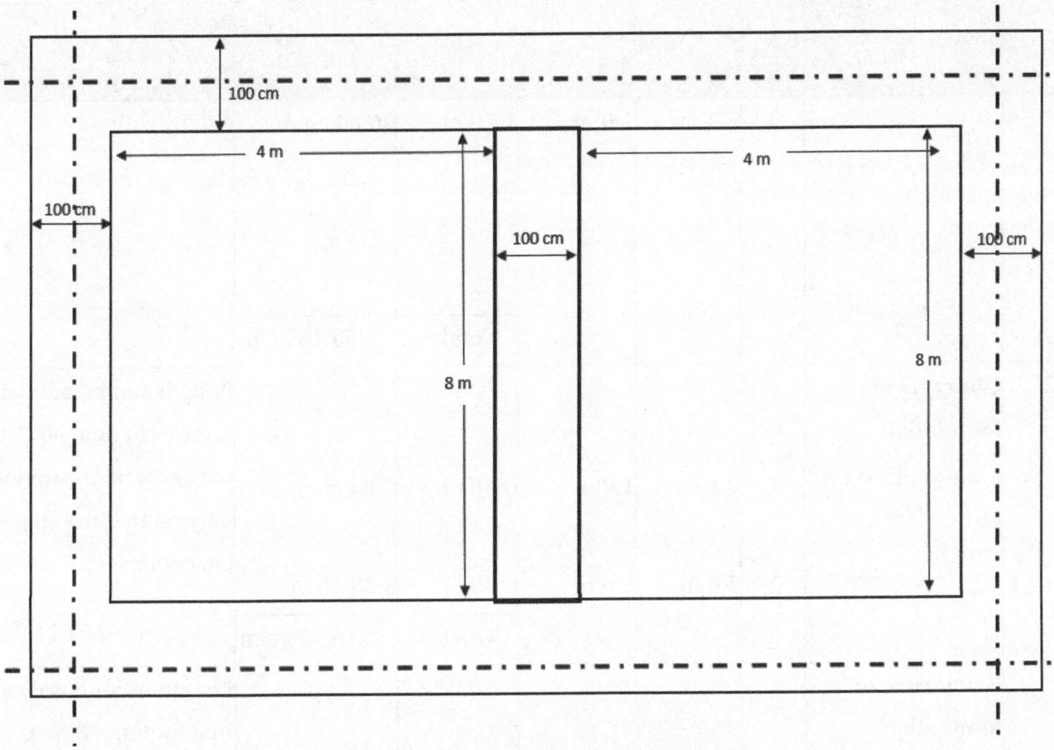

Figure 3.16 (c): Plan for earthwork

- Length of the long wall for the foundation earthwork= Centre to centre length + width of one wall

  = 8.8+ 1.0= 9.8 m (100 cm= 1m)

- Length of the short wall for foundation = Centre to centre length - width of one wall

  = 8.4-1.0= 7.4 m

- The breadth of the earthwork is 100 cm or 1 m. The excavation is done up to ground level. So, the depth for earthwork calculation is 120 cm. The rest of the calculations are shown in the Table. The same procedure is adopted for the rest of the calculations.

Table 3.12: Estimation for one-roomed building

Item no.	Description of item	No	Length	Breadth	Height	Quantity	Explanation
1	**Earthwork in excavation** Long walls Short walls	2	9.8 m	1.0 m	1.20 m	23.52 cu m	Length long wall= c/c length+ breadth of the wall= 8.8+ 1.0 = 9.8 m Length of short wall= 8.4-

Item no.	Description of item	No	Length	Breadth	Height	Quantity	Explanation
		3	7.4 m	1.0 m	1.20 m	26.64 cu m	1.0 = 7.4 m
					Total =	50.16 cu m	
2	**Concrete in foundation** Long walls Short walls	2	9.8 m	1.0 m	0.40 m	7.84 cu m	Length and breadth are same as in item no. 1 but the concrete is provided only on the first step of 40 cm depth.
		3	7.4 m	1.0 m	0.40 m	8.88 cu m	
					Total =	16.72 cu m	
3	**Brickwork in foundation** **Long walls** First Footing Second footing Plinth walls **Short walls** First Footing Second footing Plinth walls						Length of long wall for first footing= 8.8+ 0.70 = 9.5 m
		2	9.5 m	0.70 m	0.40 m	5.32 cu m	Length of short wall for first footing = 8.4- 0.7 = 7.7 m
		2	9.4 m	0.60 m	0.40 m	4.512 cu m	Length of long wall for second footing= 8.8+ 0.60 = 9.4 m
		2	9.3 m	0.50 m	0.80 m	7.44 cu m	Length of short wall for second footing = 8.4- 0.6 = 7.8 m
		3	7.7 m	0.70 m	0.40 m	6.468 cu m	
		3	7.8 m	0.60 m	0.40 m	5.616 cu m	Length of long wall for plinth= 8.8+ 0.50 = 9.3 m
		3	7.9 m	0.50 m	0.80 m	9.48 cu m	

Item no.	Description of item	No	Length	Breadth	Height	Quantity	Explanation
					Total =	38.836 cu m	Length of short wall for plinth = 8.4- 0.5 = 7.9 m
4	**Brickwork in superstructure** Long walls Short walls	2 3	9.2 m 8.0 m	0.40 m 0.40 m	4.0 m 4.0 m	29.44 cu m 38.40 cu m	Length of long wall for spuperstructure= 8.8+ 0.40 = 9.20 m Length of short wall for first footing = 8.4- 0.40=
					Total= 67.84 cu m		8.0 m
					2.5 m	3.0 cu m	
	Deductions Openings of doors	2	1.5 m	0.40 m	2.0 m	2.4 cu m	Bearing for the lintels is assumed to be 20 cm on each side.
	Opening of windows	2	1.5 m	0.40 m	0.20 m	0.304 cu m	
	Lintel over doors	2	1.9 m	0.40 m	0.20 m	0.304 cu m	
	Lintels over windows	2	1.9 m	0.40 m	Total= 6.008 cu m		
	Total brickwork= 67.84- 6.008= **61.832 cu m**						
5	**Flooring** (40 mm thick cement concrete over 10 cm concrete layer over 10 cm sand filling)						Sand filling, concrete, and cement concrete are calculated in sq m. The thickness is already given.

Item no.	Description of item	No	Length	Breadth	Height	Quantity	Explanation
	Room 1	1	8.5 m	4.5 m	-	38.25 sq m	
	Room 2	1	8.5 m	4.5 m	-	38.25 sq m	
					Total= 76.5 sq m		
6	**Roofing** (14 cm mud layer over 12 mm plaster over 2 coats of bitumen over tiles over battens)		1	8.7 m	8.3 m	72.21 sq m	Bearing is taken 10 cm

Problem 4:

The plan and section of a two-roomed building are shown in the figure shown on next page. The base of 40 cm is provided with lime concrete. Prepare an estimate for the shown two-roomed building using the centre line method with the following quantities:

- Earthwork of foundation
- Lime concrete in foundation
- Brickwork in foundation and superstructure

The size of the door is 1.5 m × 2.5 m and the size of the windows is 1.5 m × 2.0 m.

Solution:

- Centre to centre length of the long wall= $4+4+0.40+\frac{1}{2}\times 0.40+\frac{1}{2}\times 0.40= 8.80$ m
- Centre to center length of the outer short wall= $8+\frac{1}{2}\times 0.40+\frac{1}{2}\times 0.40= 8.40$ m
- Total c/c length of the building= 2×length of long wall + 3×length of short wall

  = 2×8.80+ 3×8.40= 42.8 m

Very Important Note: It should be noted that the estimation of the shadowed part is overestimated in the total wall length. Now, subtract the breadth of one wall for every component in the case of the cross wall.

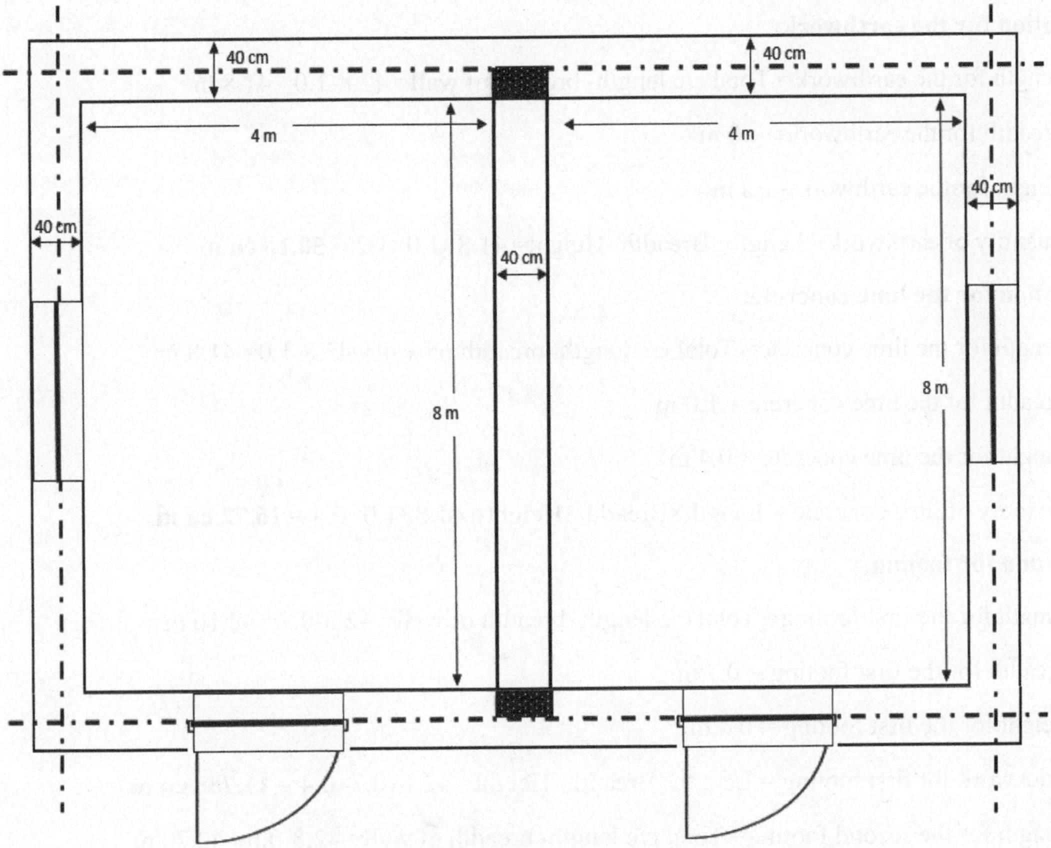

Figure 3.17 (a) Plan of the building

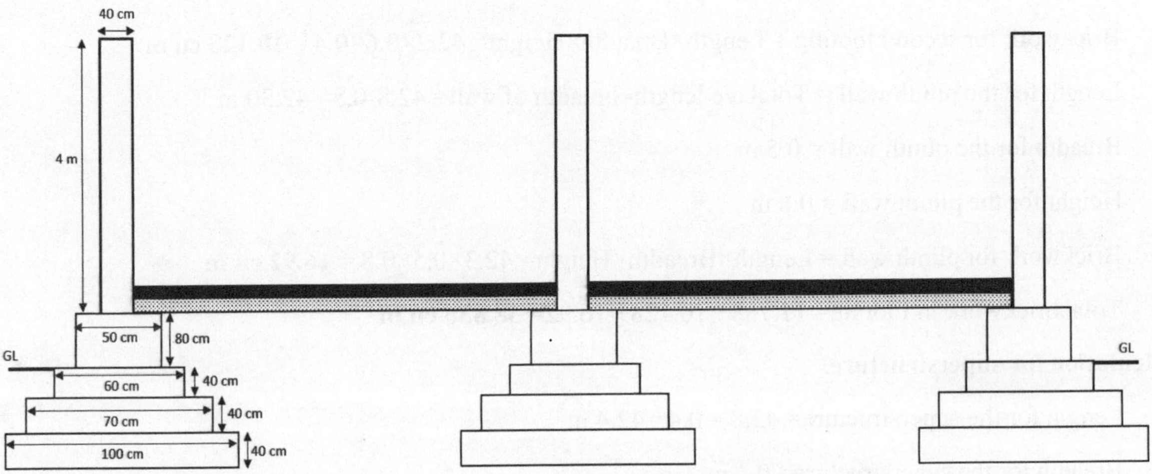

Figure 3.17 (b): Section of the building

Calculation for the earthwork:

- Length for the earthwork= Total c/c length- breadth of wall= 42.8-1.0= 41.8 m
- Breadth for the earthwork= 1.0 m
- Height for the earthwork= 1.2 m
- Quantity of earthwork= Length×Breadth×Height= 41.8×1.0×1.2 = **50.16 cu m**

Calculation for the lime concrete:

- Length for the lime concrete= Total c/c length- breadth of wall= 42.8-1.0= 41.8 m
- Breadth for the lime concrete = 1.0 m
- Height for the lime concrete = 0.4 m
- Quantity of lime concrete = Length×Breadth×Height= 41.8×1.0×0.4 = **16.72 cu m**

Calculation for footing:

- Length for the first footing= Total c/c length- breadth of wall= 42.8-0.7= 42.10 m
- Breadth for the first footing = 0.7 m
- Height for the first footing = 0.4 m
- Brickwork for first footing = Length×Breadth×Height= 42.1×0.7×0.4 = **11.788 cu m**
- Length for the second footing= Total c/c length- breadth of wall= 42.8-0.6= 42.20 m
- Breadth for the second footing = 0.6 m
- Height for the second footing = 0.4 m
- Brickwork for second footing = Length×Breadth×Height= 42.2×0.6×0.4 = **10.128 cu m**
- Length for the plinth wall = Total c/c length- breadth of wall= 42.8-0.5= 42.30 m
- Breadth for the plinth wall = 0.5 m
- Height for the plinth wall = 0.8 m
- Brickwork for plinth wall = Length×Breadth×Height= 42.3×0.5×0.8 = **16.92 cu m**
- Total brickwork in footing= 11.788+ 10.128+ 16.92**= 38.836 cu m**

Calculation for superstructure:

- Length for the superstructure= 42.8 – 0.4= 42.4 m
- Breadth for the superstructure= 0.4 m
- Height for the superstructure= 4 m

- Quantity of brickwork= 42.4×0.4×4= **67.84 cu m**
- Deductions= **6.008 cu m** (Calculated similarly as shown in the previous numerical)
- Net quantity of brickwork in superstructure= 67.84- 6.008= **61.832 cu m**

Important note: The calculations in the long wall and short wall method are relatively clear and creates no confusion in any case. However, in the case of the centre line method implemented on the structure with cross wall should be considered carefully as implemented in the given example.

UNIT SUMMARY

- A detailed estimate uses detailed designs and plans to determine the quantities of the different materials and their cost in the construction.
- Detailed information about the design, material, and their cost is required for the preparation of the detailed estimate.
- Goods and Service Tax was introduced in 2014 and implemented in 2016 in India. It is a tax that is levied on services and goods.
- The contingency is the fund/money taken as the percentage of the cost of construction that is kept to cover the unexpected costs.
- The agency charges are the internal cost of the construction company incurred due to the administrative and managerial operations.
- The whole work is divided into the items of the work in the unit quantity method.
- In the total quantity method, the items of work are divided into five categories, and the cost is calculated for these categories.
- The length of the long wall and short wall is given in the following equations too:

  $Length\ of\ long\ wall = center\ to\ center\ distance + breadth\ of\ wall$

  $Length\ of\ short\ wall = center\ to\ center\ distance - breadth\ of\ wall$
- In the centre line method, the case of cross walls in construction should be considered carefully as there may be chances of overestimation due to errors in the calculations.
- A bar bending schedule provides information about the bars that are used in the construction. A BBS is presented in the tabular form.
- Weight per metre of the bar is taken $\frac{\Phi^2}{162}$. Here, Φ is the diameter of the bar.

- The contingencies may be defined as the funds that are kept to cover the cost of uncertain events or circumstances that can occur during project implementation.
- The cost of contingencies is taken as a percentage of the cost of construction.
- The work-charged staff are the temporary employees that may be hired to perform the particular tasks.
- The cost of work-charged establishment is taken from 1.5% to 2% according to the Public Works Department (PWD).
- There is a provision to keep 8% of the estimated cost of construction for the water supply and sanitary charges.
- The prime cost refers to the actual cost of the item paid by the contractor to the merchant. The cost of carrying or installation is not included.
- The bill of quantities provides the quantities measured from the drawings which can be used to calculate the cost of construction.
- The price of equipment and machinery used at the different stages of the construction is taken as a percentage of the cost of material.
- The bill of quantities includes the cost of skilled and unskilled labor. Payment to the labor is made based on the rate schedule.

EXERCISE

Multiple Choice Questions

1. Which of the following is required for preparing the detailed estimate?

   (a) Specifications (b) Drawings

   (c) Schedule of rates (d) All of the above

2. What does GST stand for?

   (a) Gross service tax (b) Gross slab tax

   (c) Goods and services tax (d) None of the above

3. In which year, the GST is implemented in India?

   (a) 2010 (b) 2012 (c) 2016 (d) 2009

4. What percentage of the construction cost is taken as contingencies in a construction project?

   (a) 3% - 5% (b) 4% - 8%

   (c) 6% - 8% (d) 6% -10%

5. Centre to centre length of a wall is 10 m and the breadth of the wall is 40 cm. What will be the length of the long wall?

(a) 9.60 m (b) 10.40 m

(c) 9.20 m (d) 10.80 m

(Hint: $\boldsymbol{Length\ of\ long\ wall = center\ to\ center\ distance + breadth\ of\ wall}$ **)**

6. The centre to centre length of a wall is 10 m and the breadth of the wall is 40 cm. What will be the length of the short wall?

(a) 9.60 m (b) 10.40 m

(c) 9.20 m (d) 10.80 m

(Hint: $\boldsymbol{Length\ of\ short\ wall = center\ to\ center\ distance - breadth\ of\ wall}$)

7. If the diameter of the bar is Φ, what will be the length of the bend of 90°?

(a) Φ (b) 2×Φ

(c) 3×Φ (d) 4×Φ

8. If the diameter of the bar is 16 mm, what will be the length of the bend of 45°?

(a) 16 mm (b) 32 mm

(c) 48 mm (d) 64 mm

(Hint: For a 45° bend, the diameter of the bar= length of the bend)

9. Overlapping length in bars is kept:

(a) 30 Φ- 35 Φ (b) 35 Φ- 40 Φ

(c) 40 Φ – 45 Φ (d) 55 Φ – 60 Φ

10. What will the weight per metre (in Kg/m) for a bar of 12 mm?

(a) 0.84 (b) 0.87

(c) 0.89 (d) 0.91

(Hint : Weight per metre of bar= $\boldsymbol{\frac{\Phi^2}{162}}$**)**

11. If the clear span of a beam is 5500 mm and the spacing for the stirrups is taken 150 mm. The number of stirrups required will be:

(a) 37 (b) 38

(c) 39 (d) 40

(Hint: Number of stirrups required = $\boldsymbol{\frac{clear\ span\ of\ beam}{spacing} + 1}$**)**

12. According to PWD norms, what is the percentage of the work charged establishment in a construction project?
    (a) Below 1% (b) 1.5 % - 2%
    (c) 3% - 5% (d) More than 5%
13. The contractor is not allowed profit on the articles of :
    (a) Provision sum (b) Provisional items
    (c) Prime cost (d) Specialized items
14. The length to breath ratio for a one-way slab is:
    (a) Less than 1 (b) Less than 2
    (c) Equal to or more than 2 (d) None of the above
15. The length of a hook is taken:
    (a) 5Φ (b) 9Φ
    (c) 12Φ (d) 15Φ

Answers of Multiple-Choice Questions

1. (d), 2. (c), 3. (c), 4. (c), 5. (b), 6. (a), 7. (b), 8. (a), 9. (c), 10. (c), 11. (b), 12. (b), 13. (c), 14. (c), 15 (b)

Short and Long Answer Type Questions

1. Enlist the purposes of detailed estimation.
2. Write short notes on the following:
   (a) Specifications
   (b) Schedule of rates
   (c) Plans and drawings
3. Explain the procedure of preparation of the detailed estimate.
4. Explain the following briefly:
   (a) Unit quantity method
   (b) Total quantity method
5. What do you understand by a bar bending schedule? Explain the purpose of the bar bending schedules.
6. Draw and explain the different shapes of the bars.
7. Draw a bar bending schedule for an isolated footing given below:

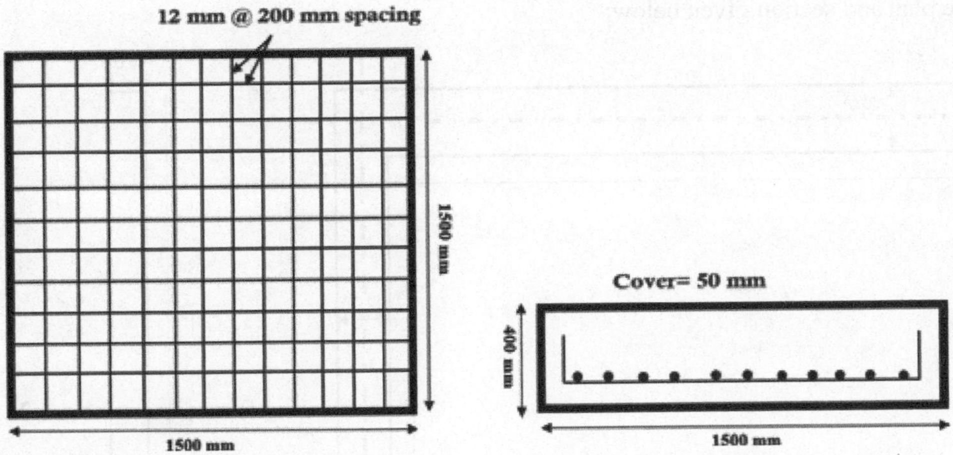

8. Write a short note on the following:
   (a) Work charged establishment
   (b) Contingencies
   (c) Spot items
   (d) Provisional sum
   (e) Provisional quantities
9. What do you understand by bill of quantities? Explain the significance of the bill of quantities. Draw the format of the bill of quantities.
10. Prepare a bar bending schedule for a slab shown as follows:

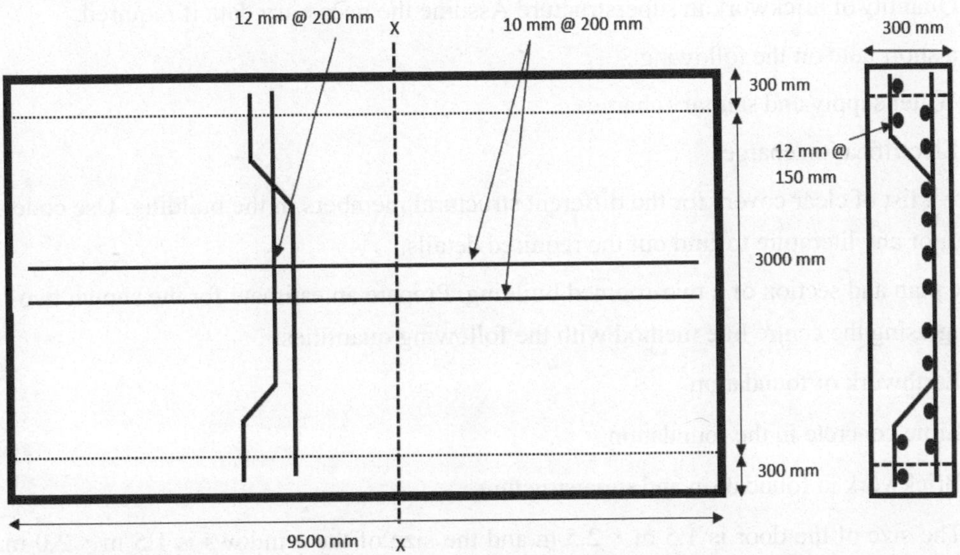

Assume the necessary data and calculate the total weight of steel bars used in this slab.

11. See the plan and section given below:

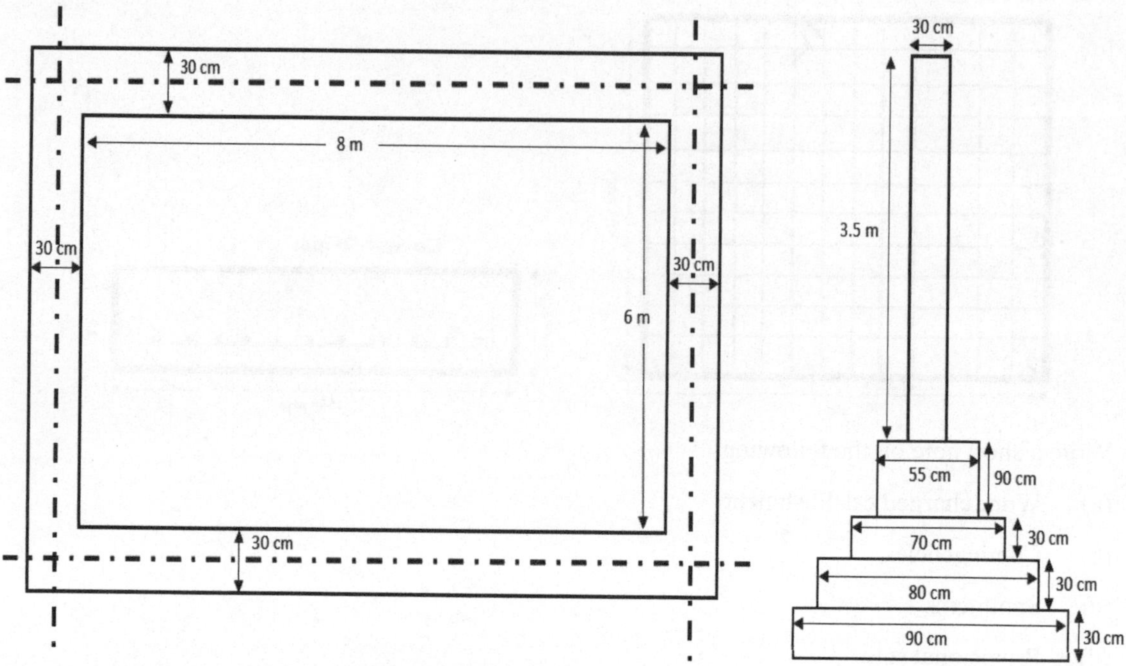

Prepare a detailed estimate of the following quantities:

(a) Earthwork assuming that the ground level is at 90 cm

(b) Quantity of lime concrete in foundation

(c) Quantity of the brickwork in the foundation

(d) Quantity of brickwork in superstructure Assume the necessary data if required.

12. Write a short note on the following:

a. Water supply and sanitary charges

b. Electrification charges

13. Prepare a list of clear covers for the different structural members in the building. Use codes, internet sources, or any literature to find out the required details.

14. See the plan and section of a two-roomed building. Prepare an estimate for the shown two- roomed building using the centre line method with the following quantities:

- Earthwork of foundation
- Lime concrete in the foundation
- Brickwork in foundation and superstructure
- The size of the door is 1.5 m × 2.5 m and the size of the windows is 1.5 m × 2.0 m. Assume the necessary data if required.

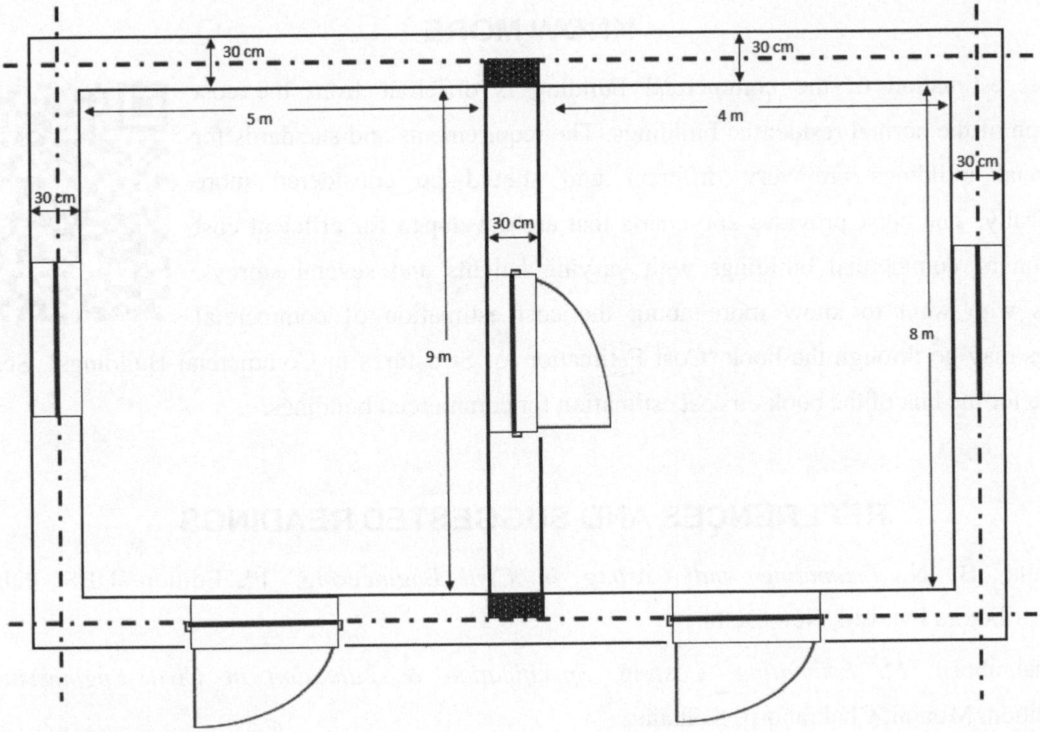

(a) Plan of the building

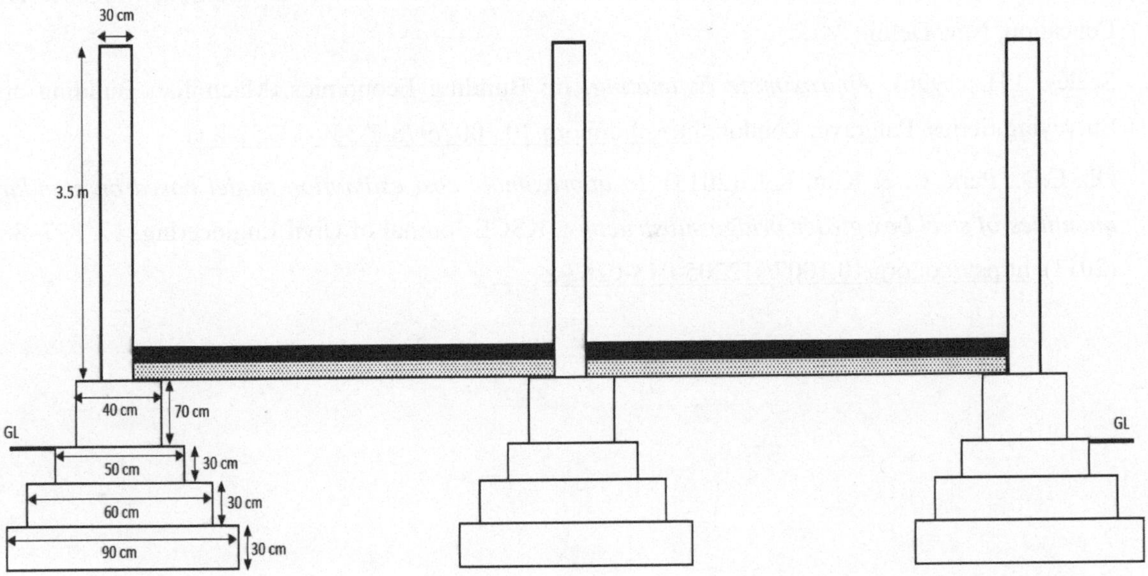

(b) Section of the building

KNOW MORE

The cost estimation of the commercial building is different from the cost estimation of the normal residential buildings. The requirements and standards for commercial buildings are very different and should be considered more scientifically. The book provides 260 charts that are developed for efficient cost estimation of commercial buildings with varying heights and several storeys. Students who want to know more about the cost estimation of commercial buildings may go through the book 'Cost Estimation of Structures in Commercial Buildings'. Scan the QR code for the link of the book on cost estimation for commercial buildings.

REFERENCES AND SUGGESTED READINGS

1. Datta, B. N. *Estimating and Costing in Civil Engineering,* 1st Edition UBS Publisher Distributors Pvt. Ltd, New Delhi
2. Chakraborti, *M. Estimating, Costing, Specification & Valuation in Civil Engineering,* 1st Edition, Monojit Chakraborti, Kolkata
3. Peurifoy,R. L. & Oberlender, D. G. *Estimating Construction Costs*, 5th Edition, McGraw Hill Education, New Delhi.
4. Seeley, I.H. (1996). *Approximate Estimating*. In: Building Economics. Macmillan Building and Surveying Series. Palgrave, London. https://doi.org/10.1007/978-1-349-13757-2_6
5. Oh, C.D., Park, C. & Kim, K.J. (2013) *An approximate cost estimation model based on standard quantities of steel box girder bridge substructure*. KSCE Journal of Civil Engineering, 17, 877–885 (2013). https://doi.org/10.1007/s12205-013-0287-z

4 Estimates for Civil Engineering Works

UNIT SPECIFICS

Through this unit, we will discuss the following aspects:

- Calculation of earthwork for roads and canals
- Application of mid-sectional area, mean sectional area, and trapezoidal formula for calculation of quantities of earthwork
- Detailed estimate of septic tank and community well
- Software and computer programs for the preparation of detailed estimates

The different methods for estimating earthwork are discussed in detail with practical examples. The detailed estimates for the septic tank and community well are elaborated on with sketches and tables. An introduction to the software used for estimation is also discussed to enhance the advanced knowledge.

This unit consists of exercises with long and short-term questions, multiple-choice questions, a list of references, and suggested reading to make the students learn through practice. Some extra information that may interest the students is provided in the form of QR codes. The students can scan these QR codes to reach the source of further knowledge on the different topics.

RATIONALE

This unit on estimates for civil engineering works presents the different techniques for the calculation of the earthwork. The methods like the mid-sectional area method, mean sectional area method and trapezoidal formula are discussed in detail. The applications of these methods for the calculation of earthwork for the roads and canals are presented using the practical problem. The detailed estimates of the septic tank and community well are discussed with solved numerical examples. The different software that are used for preparing the detailed estimates are also introduced at the end of the chapter. The content covered in this unit enhances the knowledge of students in the direction of estimates of specialized construction facilities like wells, highways, septic tanks, and canals.

After learning the detailed estimates for buildings, there is a need to understand the calculation of earthwork for linear construction like highways and canals. The detailed estimates of the septic tank and community well are different from the estimates of buildings and need the attention of the students. The earthwork calculations and estimates of septic tanks are essential to understand for a civil engineering student. Knowledge obtained by this chapter will help the students in their civil engineering career and they will be able to calculate the detailed probable cost of septic tanks and community wells. The students will also learn about soft computing and the role of software in preparing detailed estimates in construction.

PRE-REQUISITES

Knowledge of the detailed estimate and their formats is required to understand this unit.

UNIT OUTCOMES

The list of outcomes of this unit is as follows:

U4-O1: Calculate the quantities of the earthwork for highways and canals

U4-O2: Prepare the detailed estimate of the septic tank and community well

U4-O3: Understand the role of computer software and programs in construction cost estimation

Unit-4 Outcomes	EXPECTED MAPPING WITH COURSE OUTCOMES (1- Weak Correlation; 2- Medium correlation; 3- Strong Correlation)				
	CO-1	CO-2	CO-3	CO-4	CO-5
U4-O1	-	-	2	-	-
U4-O2	-	-	3	-	-
U4-O3	-	-	-	3	-

4.1 ESTIMATE FOR CIVIL ENGINEERING WORKS

Civil engineers should have a deep knowledge of estimating the cost of civil engineering works like canals, roads, reservoirs, etc. The cost estimation of the earthwork is very important to understand for the preparation of an accurate detailed estimate. The quantity of the earthwork is calculated based on the different techniques which should be understood carefully. Similarly, the estimation for the septic tank and community well is different from the estimation for buildings. In this unit, a detailed discussion of these topics is provided.

4.2 EARTHWORK

Earthowork may be defined as the removal, placement, redistribution, or compaction of the excavated materials on the construction site. The construction of structures like roads, bridges, tunnels, etc. requires modification of the landscape. So, in this case, there is a need for the movement and placement of the soil that leads to the earthwork activities. The earthwork includes the excavation, cutting, filling, grading, transporting, and compacting of the soil. Some of the important terminologies in earthwork are listed as follows:

(1) **Cutting:** The excavation or digging of the earth is called 'cutting'.

(2) **Filling:** The deposition of the earth at a low-lying place is known as 'filling'.

(3) **Banking:** The deposition of the earth above the ground level is known as 'banking'.

(4) **Lead:** Lead may be defined as the horizontal distance from the centre of gravity of the excavation of the earth to the centre of gravity of deposition of the earth.

(5) **Lift:** Lift may be defined as the vertical distance from the centre of gravity of excavation of the earth to the centre of gravity of deposition of the earth. Figure 4.1 shows the diagram of lead and lift.

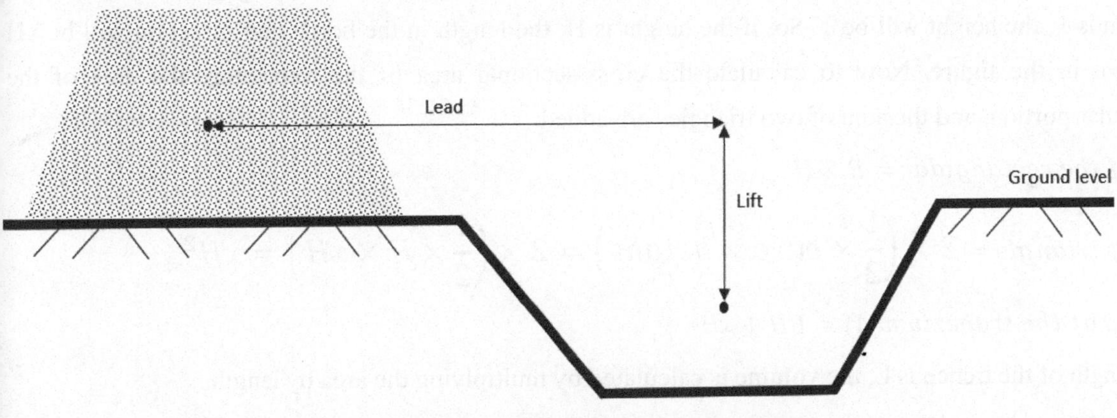

Figure 4.1: Concept of lead and lift

4.3 BASICS OF CALCULATION OF VOLUME

The cutting of the earth forms the trench. The shapes of the trenches may be rectangular or trapezoidal. The volume of the earthwork is calculated by multiplying the cross-sectional area by the length of the trench. The cross-section area of a rectangular trench can be calculated by multiplying the breadth by the depth of the trench. A typical section of the trapezoidal section is shown as follows:

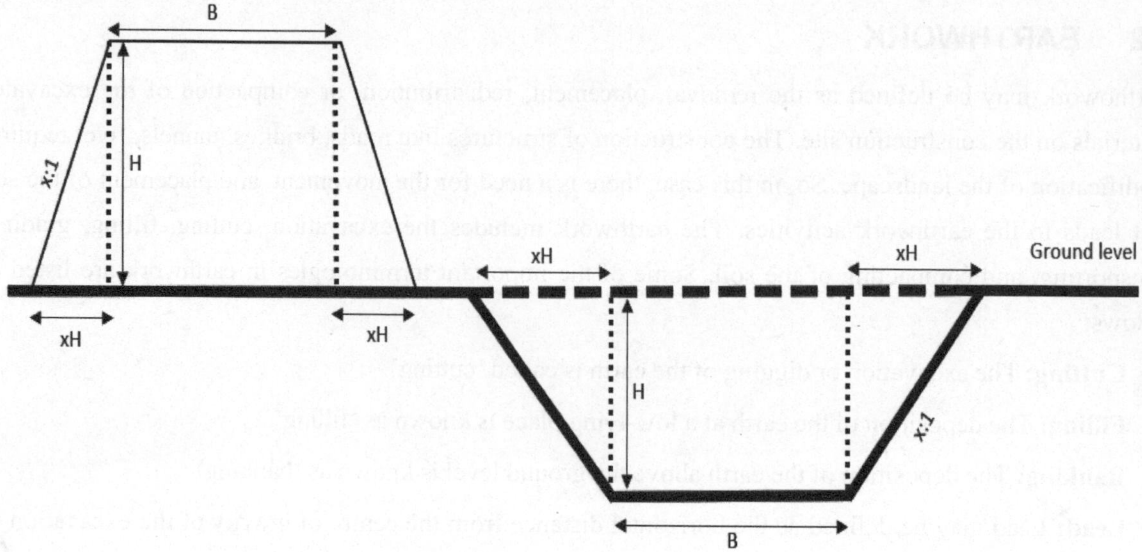

Figure 4.2: Cutting and filling

The trapezoidal-shaped cutting and filling is shown in Figure 4.2. The breadth is B and the height of the trench is H. The horizontal to the vertical slope is x:1. It means that if the length in the horizontal direction is x, the height will be 1. So, if the height is H, the length in the horizontal direction will be xH as shown in the figure. Now to calculate the cross-sectional area of the trapezium, the area of the rectangular portion, and the area of two triangles are added.

$$Area\ of\ the\ rectangular = B \times H$$

$$Area\ of\ triangle = 2 \times \left(\frac{1}{2} \times base \times height\right) = 2 \times \left(\frac{1}{2} \times H \times xH\right) = xH^2$$

$$So, Area\ of\ the\ trapezium(A) = BH + xH^2$$

If the length of the trench is L, the volume is calculated by multiplying the area by length.

$$Volume\ (V) = Area\ of\ trapezium \times Length\ of\ the\ trench = A \times L$$

However, the section may not be uniform throughout the length of the trench. The sections at the front, rear, and intermediate sections may vary in dimensions. In this case, the average of the average values of the parameter like height may be taken. Different approaches are used for calculation of the sectional area of the trenches and such sections.

4.4 METHODS OF CALCULATION OF CROSS-SECTIONAL AREA

As discussed in the previous section, the cross-section area throughout the length of the trench or embankment may or may not be uniform. So, there is a need to calculate the cross-section area using different techniques for accurate estimation. The different methods for calculation of the sectional area are listed as follows:

(1) Mid-sectional area method

(2) Mean sectional area method

(3) Prismoidal formula

(4) Trapezoidal formaula

4.4.1 Mid-sectional Area Method

In the mide-sectional area method, the mean height of the section is considered for calculation. The rest of the calculations are done as discussed in the section 4.3. The figure of the trapezium is shown below:

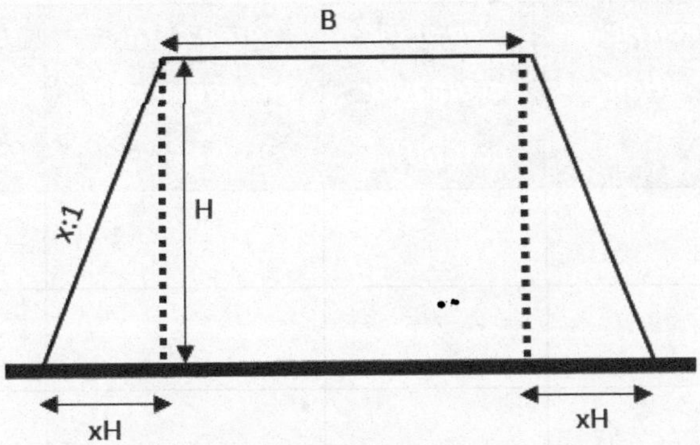

Figure 4.3 (a) : Calculations for mid-sectional area method

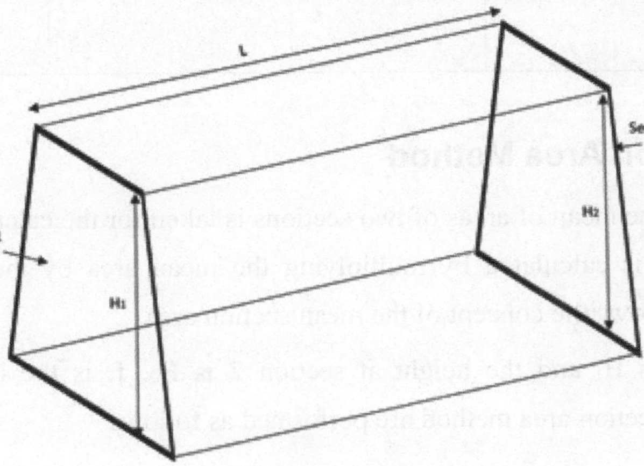

Figure 4.3(b) Isometric view of the trench

Let's assume:

L: length of the trench or embankment

H_1: Height of the trapezium at section 1

H_2: Height of the trapezium at section 2

$$Mean\ Height(Hm) = \frac{(H_1 + H_2)}{2}$$

$$Area\ of\ the\ mid-section(A_{mid}) = BH_m + xH^2{}_m$$

$$Volume\ of\ the\ earthwork = Area\ of\ mid-section \times length\ of\ trench = (BH_m + xH^2{}_m) \times L$$

The trench or embankment under consideration can be divided into multiple sections and the calculations are done in tabular form. The format for calculations in the mid-sectional area method is given in Table 4.1.

Table 4.1: Mid-sectional area method

Station	Height (H)	Mean Height (Hm)	Area of the rectangular part (BH_m)	Area of the triangular part (xH_m^2)	Total Area (A_{mid})	Length (L)	Volume

4.4.2 Mean Section Area Method

In the mean section area, the mean of areas of two sections is taken for the calculation of the volume. The volume of the earthwork is calculated by multiplying the mean area by the length of the trench or embankment. Figure 4.4 shows the concept of the mean section area.

The height at section 1 is H_1 and the height at section 2 is H_2. L is the length of the trench. The calculations for the mean section area method are performed as follows:

$$Cross-sectional\ area\ at\ section\ 1\ (A_1) = BH_1 + xH^2{}_2$$

$$Cross-sectional\ area\ at\ section\ 1\ (A_2) = BH_2 + xH^2{}_2$$

$$Mean\ sectional\ area\ (A_m) = \frac{A_1 + A_2}{2}$$

$$Volume\ of\ the\ earthwork\ \ (V)\ = \left(\frac{A_1 + A_2}{2}\right) \times L$$

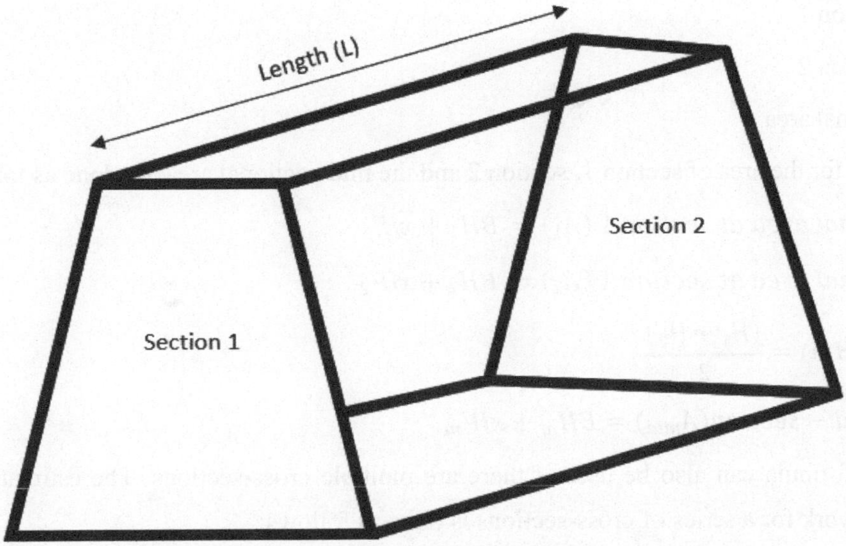

Figure 4.4: Calculations for mean sectional area method

The calculations are done in tabular form as discussed in the case of mean sectional area. Table 4.2 shows the format of the mean sectional area method.

Table 4.2: Estimation of volume using mean sectional area method

Station	*Height (H)*	*Area of the rectangular part (BH_m)*	*Area of the triangular part (xH_m^2)*	*Total Area*	*Mean sectional area (A_m)*	*Length (L)*	*Volume (A_mL)*

4.4.3 Prismoidal Formula

The prismoidal formula uses the concept of combining the mean-sectional area and the mid-sectional area. The volume using the prismoidal formula is calculated using the following equation:

$$Volume\ (V) = (A_1 + A_2 + 4A_{mid}) \times \frac{L}{6}$$

Here:

L: Length of the trench

A_1: Area of section 1

A_2: Area of section 2

A_{mid}: Mid-sectional area

The calculations for the area of section 1, section 2 and the mid-sectional area are done as follows:

$$Cross-sectional\ area\ at\ section\ 1\ (A_1) = BH_1 + xH^2{}_1$$

$$Cross-sectional\ area\ at\ section\ 1\ (A_2) = BH_2 + xH^2{}_2$$

$$Mean\ Height\ (Hm) = \frac{(H_1 + H_2)}{2}$$

$$Area\ of\ the\ mid-section(A_{mid}) = BH_m + xH^2{}_m$$

The prismoidal formula can also be used if there are multiple cross-sections. The estimation of the volume of earthwork for a series of cross-sections is done as follows:

$$Volume\ (V) = (A_1 + 4A_2 + 2A_3 + 4A_4 + 2A_5 + --------- + A_n) \times \frac{D}{3}$$

$$Volume\ (V) = [(A_1 + A_n) + 4(A_2 + A_4 + \cdots + A_{n-1}) + 2(A_3 + A_5 + A_7 + \cdots + A_{n-2})] \times \frac{D}{3}$$

$$\boldsymbol{Volume\ (V) = \frac{D}{3}[(Area\ of\ first\ section + Area\ of\ last\ section) + 2 \times Sum\ of\ area\ of\ odd\ numbered\ sections + 4 \times Sum\ of\ area\ of\ even\ numbered\ sections]}$$

Here D is the distance between the sections. The equation given above is the generalized equation for calculating the earthwork for a series of sections using the prismoidal formula.

Important Note: The most important point to note during the implementation of the prismoidal formula for a series of sections is that the number of sections should be odd. If there are an even number of sections, the calculations are done for the odd number of sections, and the area of one section is calculated separately.

4.4.4 Trapezoidal Formula

The trapezoidal formula is used to calculate the volume for a series of sections at equal intervals. If D is the distance between two consecutive sections and the areas for different sections are A_1, A_2, A_3.............. A_n, then the volume by trapezoidal formula is calculated using the following equation:

$$Volume\ (V) = (A_1 + 2A_2 + 2A_3 + 2A_4 + -------- + A_n) \times \frac{D}{2}$$

$$Volume\ (V) = [\frac{(A_1 + A_n)}{2} + A_2 + A_3 + A_4 + -------- + A_{n-1}] \times D$$

$$\boldsymbol{Volume\ (V) = \{\frac{Area\ of\ first\ section + Area\ of\ last\ section}{2} + Area\ of\ the\ rest\ sections\} \times Distance\ between\ consecutive\ sections}$$

Important Note: The mid-sectional area method and mean sectional area methods are widely used in calculating the quantity of the earthwork. The prismoidal method is more accurate than these two methods, however, the difference between the results remains near 1% only. So, due to the simplicity of solving, the mid-sectional area and mean sectional area techniques are adopted.

4.4.5 Area of Side Slopes

The erosion of the side slopes is a common problem along the facilities like highways and canals. So, a turfing or pitching of stones is provided along the side slopes to reduce this soil erosion. The area of the side slopes is also estimated for pitching or turfing of stones. Figure 4.5 shows the concept of side slopes. From the figure, it can be seen that the area of the ABCD region is to be calculated.

Here,

d: distance between two consecutive sections H: Height of perpendicular

The slope is x:1. So, the length of the base (A'D)= xH

$$Length\ of\ hypotenuse\ (AD) = \sqrt{H^2 + (xH)^2} = H\sqrt{1 + x^2}$$

$$Area\ of\ sideslope = d(H\sqrt{1 + x^2})$$

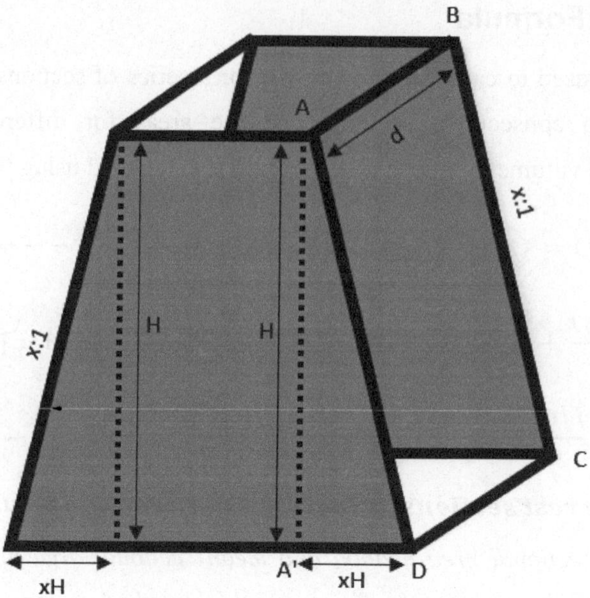

Figure 4.5: Area of side slopes

There are two side slopes. So, the area calculated in the previous step is multiplied by 2 to get the total area of side slopes. The calculations for the area of side slopes are also done in tabular format.

Table 4.3: Calculation of area of side slopes

Station	Height (H)	Mean height	Hypotenuse of side slope(Breadth of side slope) $(H\sqrt{1+x^2})$	Distance between stations (d)	Area of side slopes $2dH\sqrt{1+x^2})$

4.5 EARTHWORK QUANTITIES FOR ROADS

The calculation of the earthwork for the highways is an important task for civil engineers. The earthwork for the roads is prepared based on the longitudinal section and cross-section along the length of the road. The following points are kept in mind while preparing the estimate of the earthwork along highways:

(1) The formation level of the highway is marked. The consideration for the drainage structures should be given during marking.

(2) The formation level is kept above the ground level to avoid the accumulation of rainwater.

3) The cutting is required whenever the ground level is higher than the required formation level and filling is required when the ground level is below the designed formation level.

4) The attempt to balance the cutting and filling is made to achieve the economy.

5) For the longitudinal section, the scale of 1 cm= 10 m to 30 m is taken while for the vertical section, the scale of 1 cm= 1 m to 5 m is taken. The selection of scale depends on the slope and topographic characteristics of the terrain.

Let's try to understand the calculations for the earthwork quantities of the road with some solved examples.

Solved Example 1: Let's calculate the quantities of the earthwork for a 350 m-long road. The height at the first section is 1.5 m and height at the other section is 1.2 m. The formation width is 9 m. The side slope is given 2.5:1. Calculate the volume using the mid-sectional method, mean section method, and prismoidal method.

Solution: Given data:

- Height at section 1 (H_1)=1.5 m
- Height at section 2 (H_2)= 1.2 m
- Formation width (B)= 9 m
- Length of the road (L)= 350 m
- Side slope (x:1)= 2.5:1

Calculations for mid-sectional area method

The calculations for the mid-sectional area method are performed using the formulas discussed in section 3.4.1.

$$Mean\ Height\ (Hm) = \frac{(H_1 + H_2)}{2} = \frac{(1.5 + 1.2)}{2} = \frac{2.7}{2} = 1.35\ m$$

$$Area\ of\ the\ mid-section(A_{mid}) = BH_m + xH^2{}_m = 9 \times 1.35 + 2.5 \times 1.35^2 = 16.706 \text{ sq m}$$

$$Volume\ of\ the\ earthwork = Area\ of\ midsection \times length\ of\ trench$$

$= 16.706 \times 350 = \mathbf{5847.1\ cu\ m}$

Calculations for mean sectional area method

The calculations for the mid-sectional area method are performed using the formulas discussed in section 3.4.2.

$$Cross-sectional\ area\ at\ section\ 1\ (A_1) = BH_1 + xH^2{}_1$$

$= 9 \times 1.5 + 2.5 \times 1.5^2 = 19.125\ sq\ m$

$Cross - sectional\ area\ at\ section\ 1\ (A_2) = BH_2 + xH^2{}_2$

$= 9 \times 1.2 + 2.5 \times 1.2^2 = 14.40\ sq\ m$

$$Mean\ sectional\ area\ (A_m) = \frac{A_1 + A_2}{2} = \frac{19.125 + 14.40}{2} = 16.763\ sq\ m$$

$Volume\ of\ the\ earthwork\ (V) = Mean\ sectional\ area\ (A_m) \times L$

$= 16.763 \times 350 = \mathbf{5866.9\ cu\ m}$

Calculations for prismoidal formula

The calculations for estimation of quantities of earthwork using the prismoidal formula are performed as discussed in section 3.4.3.

$Cross - sectional\ area\ at\ section\ 1\ (A_1) = BH_1 + xH^2$

$= 9 \times 1.5 + 2.5 \times 1.5^2 = 19.125\ sq\ m$

$Cross - sectional\ area\ at\ section\ 1\ (A_2) = BH_2 + xH^2$

$= 9 \times 1.2 + 2.5 \times 1.2^2 = 14.40\ sq\ m$

$$Mean\ Height\ (Hm) = \frac{(H_1 + H_2)}{2} = \frac{(1.5 + 1.2)}{2} = \frac{2.7}{2} = 1.35\ m$$

$Area\ of\ the\ mid - section(A_{mid}) = BH_m + xH^2{}_m = 9 \times 1.35 + 2.5 \times 1.35^2 = 16.706$ sq m

$$Volume\ (V) = (A_1 + A_2 + A_{mid}) \times \frac{L}{6}$$

$$= (19.125 + 14.40 + 4 \times 16.706) \times \frac{350}{6}$$

$= \mathbf{5853.69\ cu\ m}$

Important Note: It can be seen from the calculations in the numerical given above that the results from the prismoidal method are barely different from the other two methods. The variation in the answer is not more than 1%. However, the prismoidal method is more accurate as compared to the mid section method and mecan section method. The given example is only for clearing the concept. Some more examples are given in the Solved Numerical Problems section.

4.6 EARTHWORK QUANTITIES FOR CANALS AND EMBANKMENTS

The quantity of the earthwork in the irrigation canals is calculated in the similar way as for the highways. The side slope is kept 1:1 for the cutting and varies from 1.5:1 to 2:1 for the embankment. However, the ngle of side slope depends on the type of the soil. The canals are constructed in three types of sections isted below:

1) Canal with fully cutting section
2) Canal section fully in embankment
3) Canal with partly in cutting and partly in embankment section

1) Canal with fully cutting section: The section of the canal constructed fully in cutting is shown in Figure 4.6. The sectional area of the excavated portion is calculated as discussed in section 4.3. The total sectional area is given by the following formula:

$Sectional\ area\ of\ cutting\ (A) = BH + xH^2$

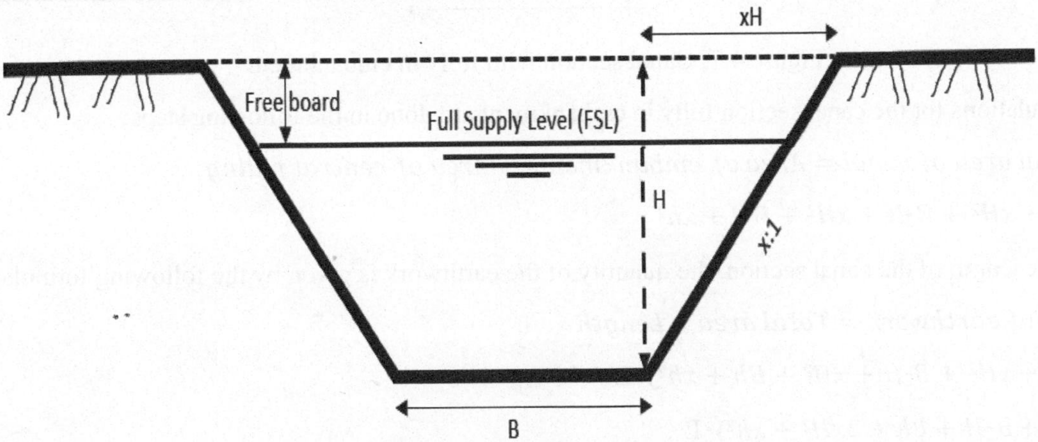

Figure 4.6: Canal with fully cutting section

If the length of the canal is L, the volume is calculated by multiplying the area by length.

$$Volume\ (V) = Sectional\ area\ of\ cutting \times Length\ of\ the\ canal = A \times L$$

The land to be occupied to construct a canal is also calculated to estimate the cost of the land acquisition. Some extra land for the construction of the service road or future development is also occupied with the land required for the canal excavation. The permanent land required for the canal project is calculated as follows:

$$Width\ of\ the\ permanent\ land\ (W) = B + 2 \times (xH) + Extra\ land$$

$$Area\ of\ permanent\ land = Width \times Length\ of\ the\ canal = W \times L$$

(2) Canal section fully in embankment

The canal section fully in embankment requires filling only. The earth is excavated from any pit and deposited for making the embankment. The canal section fully in embankment will have three parts in filling as shown in figure 4.7.

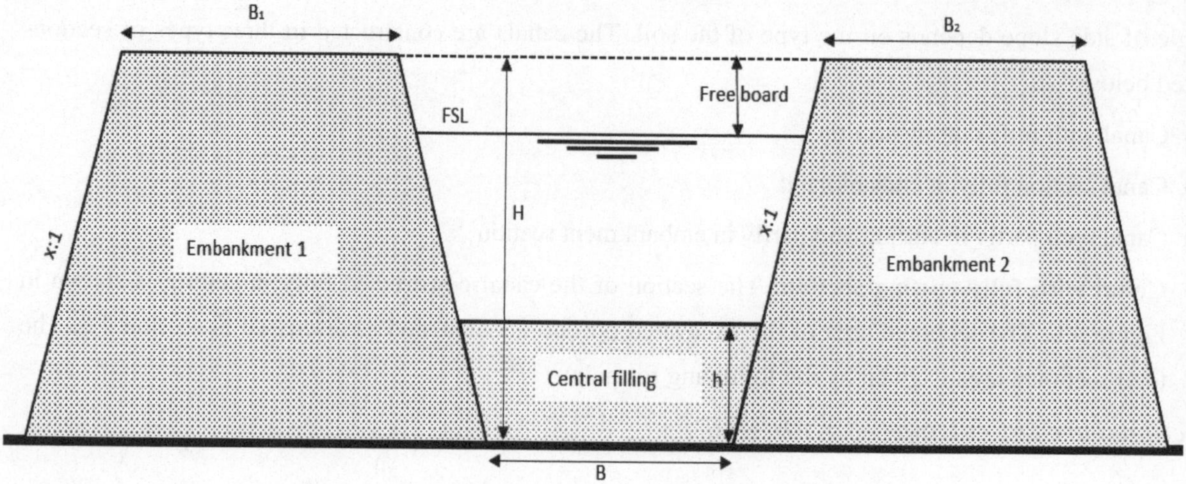

Figure 4.7: Canal section with fully in embankment

The calculations for the canal section fully in embankment are done in the following steps:

$$Sectional\ area\ of\ canal = Area\ of\ embankments + Area\ of\ central\ filling$$

$$= B_1H + xH^2 + B_2H + xH^2 + BH + xh^2$$

If L is the length of the canal section, the quantity of the earthwork is given by the following formula:

$$Volume\ of\ earthwork = Total\ area \times Length$$

$$= (B_1H + xH^2 + B_2H + xH^2 + Bh + xh^2)\times L$$

$$= (B_1H + B_2H + Bh + 2xH^2 + xh^2)\times L$$

The area of land to be occupied is also calculated as discussed in the previous section. The land area is calculated as follows:

$$Width\ of\ the\ permanent\ land\ (W) = B_1 + B_2 + B + 4(xH) + 2xh + Extra\ land$$

$$Area\ of\ land\ to\ be\ occupied = (B_1 + B_2 + B + 4(xH) + 2xh + Extra\ land) \times L$$

(3) Canal section partly in cutting and partly in filling

The canal section may be constructed partly in cutting and partly in filling as shown in figure 4.8.

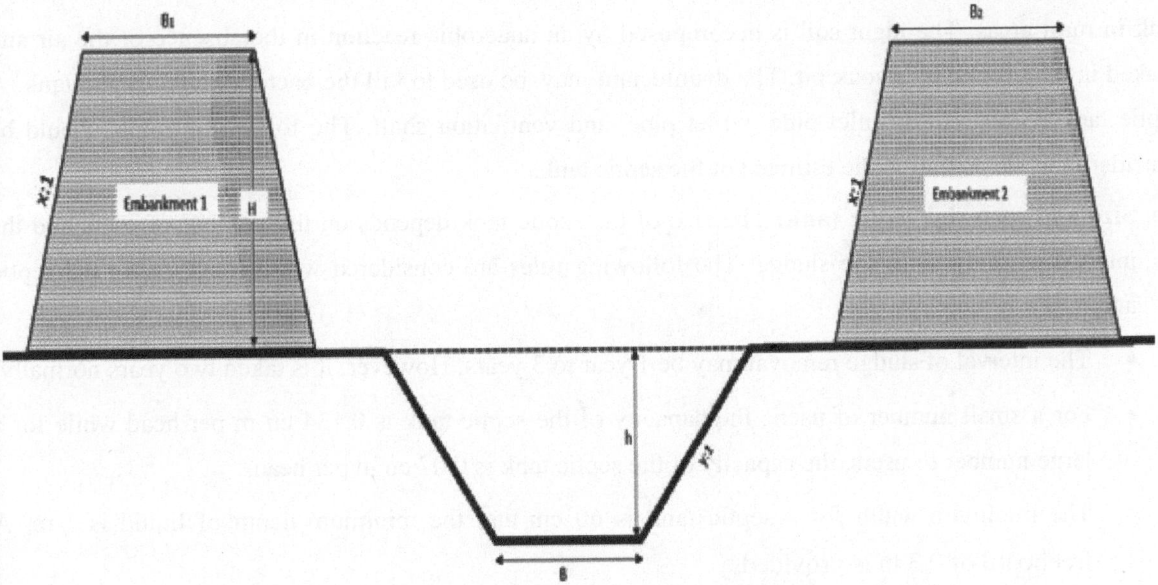

Figure 4.8: Canal section partly in cutting and partly in filling

The quantity of the earthwork for the cutting and filling is calculated separately. The calculations for the sections partly in cutting and partly in filling are discussed in the following steps:

$$Sectional\ area\ of\ filling = B_1H + xH^2 + B_2H + xH^2$$

$$= (B_1 + B_2) \times H + 2xH^2$$

$$\text{Volume in filling} = \{(B_1 + B_2) \times H + 2xH^2\} \times \text{length of section}$$

$$Sectional\ area\ of\ cutting = (Bh + yh^2) \times \text{length of section}$$

Three cases arise in the canal section partly in cutting and partly in filling discussed as follows:

- If the quantity of the cutting earthwork and filling earthwork are nearly equal, it is called 'economical digging'. In this case, the canal is designed with balancing depth.
- If the quantity of earthwork in cutting is more than required for filling, the extra soil is used for spoil banks.
- If the quantity of earthwork in cutting is less than that required for filling, the extra soil is transported from the pits. In this case, no earthwork is considered for spoil banks.

4.7 DETAILED ESTIMATE OF SEPTIC TANK

The septic tank is constructed for the disposal of the night soil in areas with no sewer facilities. The septic tanks may be constructed in the industries and rural areas. The latrine seats may be connected to the septic tank in rural areas. The night soil is decomposed by an anaerobic reaction in the absence of the air and soaked in the ground by a soak pit. The disinfectant may be used to kill the bacteria in the septic tank. A septic tank consists of an inlet pipe, outlet pipe, and ventilation shaft. The following points should be considered while preparing the estimate of the septic tank.

(1) **Size and material septic tank:** The size of the septic tank depends on the number of users and the interval of removal of the sludge. The following rules are considered while constructing the septic tank:

- The interval of sludge removal may be 1 year to 3 years, However, it is taken two years normally.
- For a small number of users, the capacity of the septic tank is 0.134 cu m per head while for a large number of users, the capacity of the septic tank is 0.07 cu m per head.
- The minimum width for a septic tank is 60 cm and the minimum depth of liquid is 1 m. A freeboard of 0.3 m is provided.
- The length-to-breadth ratio of a septic tank varies from 2 to 4.
- The septic tank is constructed using bricks or stones in cement mortar. The foundation is also constructed in cement mortar. The concrete mix of 1:3:6 or 1:4:8 may be used. The inner and outer walls are plastered with cement plaster of 1:3.

(2) **Ventilation pipe in septic tank:** A ventilation pipe must be provided with the septic tank. The minimum diameter of the septic tank is 50 mm for a septic tank. If the septic tank is situated at a distance of more than 15 m from the habitable building, the height of the ventilation pipe is carried to 1.8 m. If the septic tank is within a distance of 15 m from the habitable building, the height of the ventilation pipe should be 1.8 m above the roof of the building.

(3) **Commissioning new septic tank and desludging:** If a new septic tank is required to be in operation, it is filled with water up to the outlet level. The sludge from the other septic tank or cow dung is seeded in the septic tank to catalyze the decomposition. The removal of the sludge from the septic tank is known as desludging. The desludging is normally done in 2 years, however, the time for desludging may vary from 12 months to 36 months.

(4) **Soak pit:** The effluent from the septic tank is disposed of in the soak pit or any drain or river. However, the site for the disposal of the effluent should be at least 7.5 m away from any drinking water resource. The diameter of the soak pit should not be less than 90 cm and the minimum depth should be 1.5 m below the inverted level of the inlet pipe. The size of the soak pit may be 1 sq m per head user to 1.5 sq m per head user. The walls of the soak pit are raised above the ground to prevent water entry.

(5) **Sub-soil drains:** The sub-soil drains are provided as Stone Ware (SW) pipes. The SW pipes are laid in the trenches and surrounded by coarse aggregates.

4.8 SIZES OF SEPTIC TANKS (AS PER IS: 2470 (Part I)-1985)

Indian standard code IS: 2470 Part-I provides the details of suitable sizes of the septic tanks to be constructed for different types of buildings. The code was reaffirmed in 2001. A freeboard depth of 300 mm should be provided.

The sizes of the septic tank for a small number of users (up to 20) are given in Table 4.4 and the suitable sizes of septic tank for a residential colony are given in Table 4.5. If the students want to know more about the provisions of the IS code, they can download the IS: 2470 part I by scanning the QR code.

Table 4.4: Suitable sizes of septic tanks for a small number of users

No of users	*Length*	*Breadth*	*Liquid depth*	
			1-year cleaning interval	*2-year cleaning interval*
5	1.5 m	0.75 m	1.0 m	1.05 m
10	2.0 m	0.90 m	1.0 m	1.40 m
15	2.0 m	0.90 m	1.3 m	2.00 m
20	3.0 m	1.10 m	1.3 m	1.80 m

Table 4.5: Suitable sizes of septic tanks for residential colonies

No of users	*Length*	*Breadth*	*Liquid depth*	
			1-year cleaning interval	*2-year cleaning interval*
50	5.0 m	1.6 m	1.3 m	1.4 m
100	5.7 m	2.1 m	1.4 m	1.7 m
150	7.7 m	2.4 m	1.4 m	1.7 m
200	8.9 m	2.7 m	1.4 m	1.7 m
300	10.7 m	3.3 m	1.4 m	1.7 m

Solved Example 2: Let's prepare a detailed estimate of a septic tank as shown in figure 4.9. Assume the relevant data.

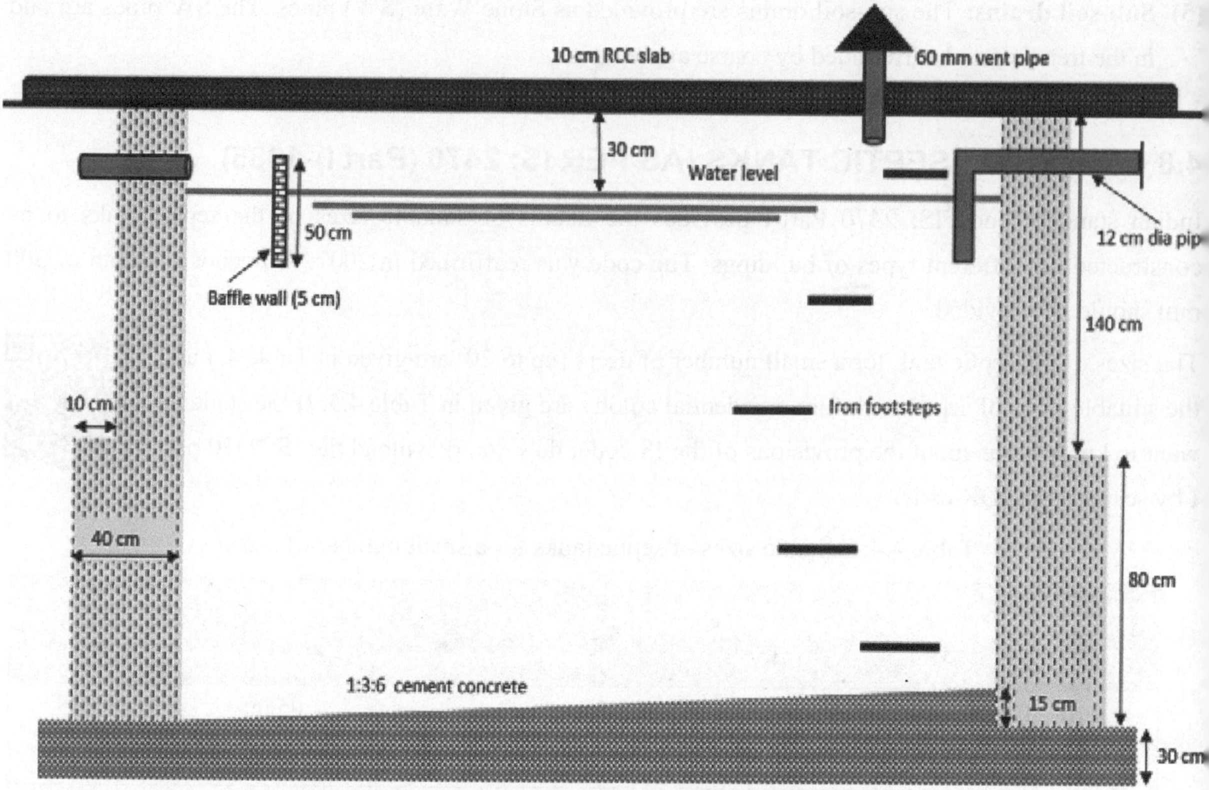

Figure 4.9 (a): Longitudinal section of septic tank

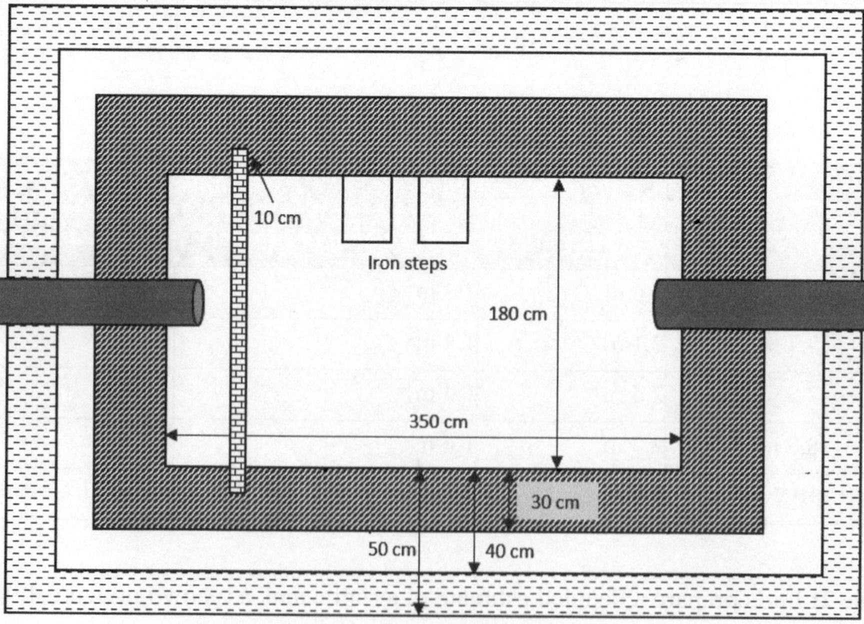

Figure 4.9 (b): Plan of soak pit

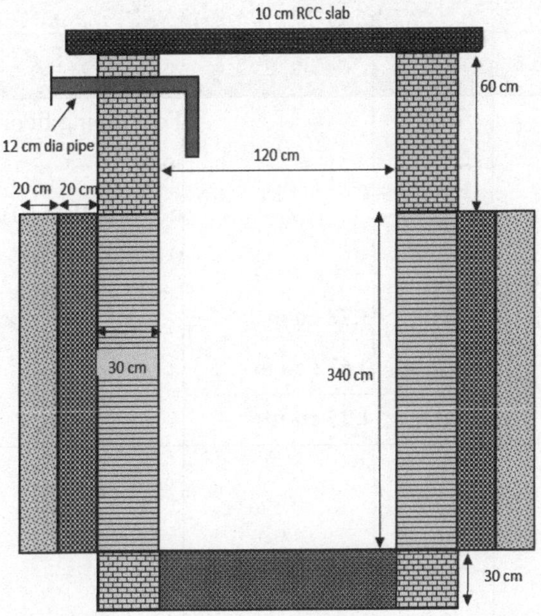

Figure 4.9 (c): Longitudinal section of soak pit

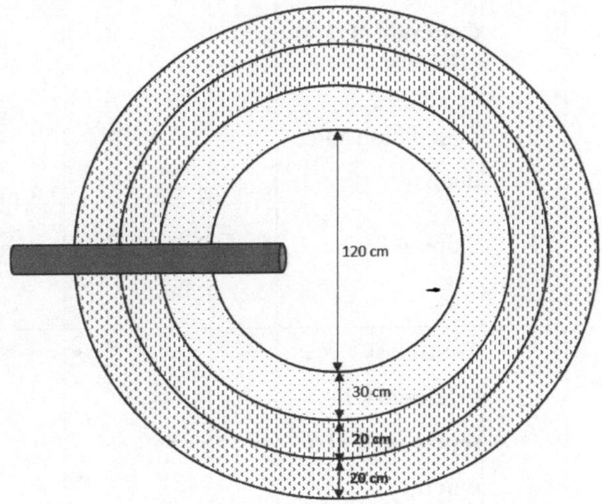

Figure 4.9(d): Plan of soak pit

Solution:

The detailed estimate for the septic tank is given in the tabular form:

Table 4.6: Taking off the quantities

Item no.	Description of item	No	Length	Breadth	Height	Quantity	Explanation
1	**Earthwork in excavation**						**From figure (b)** Plan of septic tank
	Septic tank	1	4.5 m	2.8 m	2.5 m	31.5 cu m	Length= 3.5+ 0.50+ 0.50= 4.5 m
	Soak pit up to 4 m depth	1	$\pi/4\times 2.6^2$ Sq m	-	4.0 m	21.24 cu m	Breadth= 1.8+0.50+0.50= 2.8 m
	Soak pit after 4 m depth	1	$\pi/4\times 1.8^2$ Sq m	-	0.30 m	0.76 cu m	**From Figure d:** Volume of soak pit= Area× height
					Total =	**53.5 cu m**	

Item no.	Description of item	No	Length	Breadth	Height	Quantity	Explanation
2	**Cement Concrete (1:2:4)**						For sloping floors, mean height= (0.15+0)/2 = 0.075 m
	Foundation and floor	1	4.5 m	2.8 m	0.30 m	3.78 cu m	
	Sloping floor	1	3.5 m	1.8 m	0.075 m	0.472 cu m	
					Total =	**4.25 cu m**	
3	**First-class brickwork in septic tank (1:6 mortar)**						
	Long walls Top portion Bottom portion **Short walls** Top portion Bottom portion	2	4.10 m	0.30 m	1.4 m	3.44 cu m	
		2	4.30 m	0.40 m	0.80 m	2.752 cu m	
						1.512 cu m	
		2	1.8 m	0.30 m	1.4 m		
		2	1.8 m	0.40 m	0.80 m	1.152 cu m	
					Total =	**8.856 cu m**	
4	**Second-class brickwork in soak pit**						**From Figure d:** Mean diameter= (Outer dia+ inner dia)/2 = (120+180)/2= 150 cm= 1.5 cm
	Top portion	1	π×1.5 m	0.30 m	0.6 m	0.85 cu m	
	Bottom portion	1	π×1.5 m	0.30 m	0.3 m	0.42 cu m	
					Total=	**1.27 cu m**	

Item no.	Description of item	No	Length	Breadth	Height	Quantity	Explanation
5	**Dry brick masonry in soak pit**	1	π×1.5 m	0.30 m	3.4 m	**4.81 cu m**	From figure c and d
6	**Precast RCC work**						
	Roof cover slab of septic tank	1	4.10 m	2.40 m	0.10 m	0.984 cu m	
	Roof cover slab of soak pit	1	$\pi/4\times 1.8^2$	-	0.10 m	0.254 cu m	
	Baffle wall	1	2.0 m	0.05 m	0.50 m	0.050 cu m	
					Total= 1.288 cu m		
7	**12 mm cement plaster (1:3) on the internal walls of the septic tank**						Plaster is done only on the internal walls.
	Long wall	2	3.5 m	-	2.2 m	15.4 sq m	
	Short wall	2	1.8 m	-	2.2 m	7.92 sq m	
	Floor	1	3.5 m	1.8 m	-	6.3 sq m	
					Total= 29.62 sq m		
8	**50 mm brick aggregates in soak pit**						Volume= Perimeter×breadth ×height
	Outer circle of soak pit	1	π×2.0 m	0.20 m	3.4 m	4.273 cu m	Volume= Area×height
	Floor of soak pit	1			0.3 m	0.339 cu m	

Item no.	Description of item	No	Length	Breadth	Height	Quantity	Explanation
			$\pi/4 \times 1.2^2$				
					Total= 4.612 cu m		
9	**Coarse sand layer on the outer wall**	1	π×2.40 m	0.20 m	3.0 m	**4.524 cu m**	
10	**Iron bar footsteps**	4	Bought in numbers				
11	**12 cm dia SW pipes laid in 1:3 cement mortar**						
	From latrine to septic tank	1	6 m	-	-	6 m	
	From septic tank to soakpit	1	6 m	-	-	6 m	
					Total= 12 m		
12	**50 mm ventilating pipe**	1	1.8 m	-	-	**1.8 m**	
13	**50 mm CI cowl at the top**	1	Bought in numbers				
14	**Site clearance**		Lump sum cost is added				

Table 4.7: Abstract of estimated cost

Item No.	Description or particulars of the work	Quantity	Unit	Rate	Amount
1 2 3 4 5 6	Earthwork in Excavation Cement Concrete (1:2:4) First-class brickwork in septic tank (1:6 mortar) Second-class brickwork in soak pit Dry brick masonry in soak pit Precast RCC work 12 mm cement plaster (1:3) on the internal walls of the septic tank 50 mm brick aggregates in soak pit	53.5 4.25 8.856 1.27 4.81 1.288	cu m cu m cu m cu m cu m cu m sq m cu m	80 ₹/ cu m 6200 ₹/ cu m 4393 ₹/ cu m 3200 ₹/ cu m 2300 ₹/ cu m 6000 ₹/ cu m 156 ₹/ sq m 2100 ₹/ cu m	4280 ₹ 26350 ₹ 38905 ₹ 4064 ₹ 11063 ₹ 7728 ₹
7		29.62			4621 ₹
8		4.612			9685 ₹
9 10 11 12 13 14	Coarse sand layer on the outer wall Iron bar footsteps 12 cm dia SW pipes laid in 1:3 cement mortar 50 mm ventilating pipe 50 mm CI cowl at the top Site clearance	4.524 4 12 1.8 1 -	cu m Nos m m Nos Lump sum	1600 ₹/ cu m 70 ₹/ piece 750 ₹/ m 200 ₹/ m 200 ₹/ piece 1500 ₹	7238 ₹ 280 ₹ 9000 ₹ 360 ₹ 200 ₹ 1500 ₹
Total cost					125274.00 ₹
Contingencies and work charged establishment (5%)					6263.70 ₹
Total cost					**131537.7 ₹**

The estimated cost of construction of the given septic tank is **131538 ₹.**

4.9 DETAILED ESTIMATE OF COMMUNITY WELL

The wells are still used to obtain drinking water or water for irrigation purposes. The wells used for drinking water or irrigation are known as community wells. The construction of the community wells includes the excavation, brickwork, soil filling, and ring beam work. The excavation for the wells is done in steps and the cost of the excavation depends on the type of soil. There may be a soft or hard stratum that needs to be excavated. The excavation work in the wells takes place in cutting as well as in filling.

The steps or levels in the walls have a significant impact on the quantity of the earthwork in cutting and filling. So, the steps should be considered carefully while calculating the quantities for the wells. The thickness of the walls of the well varies with the depth. Dry masonry and masonry in mortar both are used for the construction of the walls. The laterite masonry is widely used for the construction of the walls of the community wells. The estimation of the wells can be understood clearly by a solved example. An example is discussed in a detailed manner to enhance the understanding of the students.

Solved Example 3: *Let's prepare a detailed estimate of an open well as shown in Figure 10 with the following data:*

- Internal diameter of the well= 8 m
- Thickness of the brick wall= 40 cm
- Height of the brick wall= 0.75 m
- Height for excavation= 13 m

The laterite and dry laterite masonry are used in the walls as shown in Figure 4.10

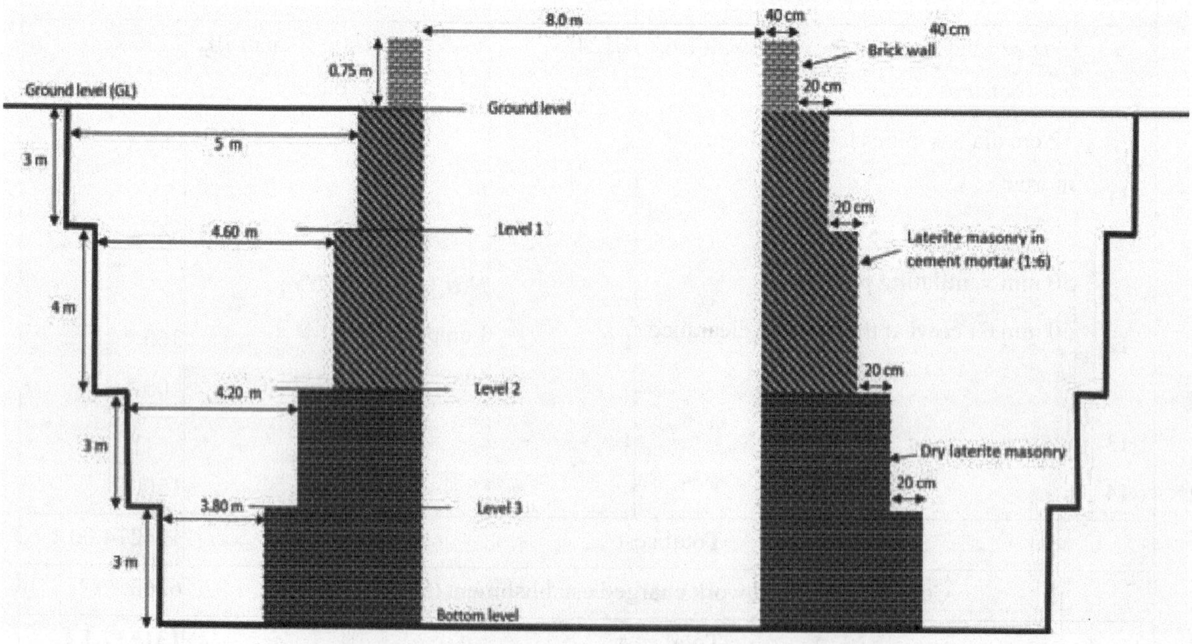

Figure 4.10: Community well

Solution: Given data:

- Internal diameter of the well= 8 m
- Thickness of the brick wall= 40 cm
- Height of the brick wall= 0.75 m
- Height for excavation= 13 m

The well is circular and made of different materials. The walls above the ground level are made of brick masonry. Laterite masonry in cement mortar is used from ground level to level 2. Dry laterite masonry is used from level 2 to the bottom level. The laterite masonry is used in the construction of the open well. From ground level to level 2 the laterite masonry in 1: 6 cement mortar is used while below level 2 dry laterite masonry is used. The detailed estimate of the open well is given in Table 4.8.

Table 4.8: Taking off the quantities

Item no.	*Description of item*	*No*	*Length*	*Breadth*	*Height*	*Quantity*	*Explanation*
1	**Earthwork in excavation**						Dia upto level 1= 5+5+8+ 0.60 +0.60 = 19.20 m
	From GL to level 1	1	$\pi/4 \times 19.20^2$		3 m	868.59 cu m	Dia between
	Level 1 to level 2						level 1 to 2=
	Level 2 to level 3	1	$\pi/4 \times 18.80^2$		4 m	1110.36 cu m	4.6+4.6+8+ 0.80 +0.80 = 18.80 m
	Level 3 to bottom level	1	$\pi/4 \times 18.40^2$		3 m	797.71 cu m	Dia between level 2 to 3=
		1	$\pi/4 \times 18^2$		3m	763.41 cu m	4.2+4.2+8+ 1+1 = 18.40 m Dia between level 3 to bottom= 3.8+3.8+8+ 1.2+1.2 = 18 m
					Total= 3540.1 cu m		

Item no.	Description of item	No	Length	Breadth	Height	Quantity	Explanation
2	**Dry laterite masonry**						External dia for level 2 to level 3= 8+1+1= 10 m
	Level 2 to level 3	1	$\pi/4\times(10^2 - 8^2)$		3 m	84.82 cu m	External dia for level 3 to bottom level= 8+1.2+1.2 = 10.4 m
	Level 3 to Bottom level	1	$\pi/4\times(10.40^2 - 8^2)$		3 m	104.05 cu m	
					Total= 188.87 cu m		
3	**Laterite masonry in 1:6 cement mortar**						External dia from GL to level 1= 8 +0.6+0.6= 9.2 m
	GL to level 1	1	$\pi/4\times(9.2^2 - 8^2)$		3 m	61.92 cu m	External dia from level 1 to level 2= 8 +0.8+0.8= 9.6 m
	Level 1 to level 2	1	$\pi/4\times(9.60^2 - 8^2)$		4 m	112.64 cu m	
					Total= 174.56 cu m		
4	**Refilling of soil**						External dia= Internal dia of well+ width of walls+ width of excavation
	GL to level 1	1	$\pi/4\times(19.2^2 - 9.2^2)$		3 m	669.16 cu m	
	Level 1 to level 2				4 m	820.84 cu m	
	Level 2 to level 3	1	$\pi/4\times(18.8^2 - 9.6^2)$		3 m	562.09 cu m	Internal dia= Internal dia of

		1	$\pi/4\times(18.4^2 - 10^2)$			well+ breadth of walls
	Level 3 to bottom level	1	$\pi/4\times(18^2 - 10.4^2)$	3 m	508.56 cu m	
				Total= 2560.65 cu m		
5	**Brickwork above ground level**	1	$\pi/4\times(8.8^2 - 8^2)$	0.75 m	**7.92 cu m**	External dia= 8+0.4+0.4= 8.8 m

Table 4.9: Abstract of estimated cost

Item No.	*Description or particulars of the work*	*Quantity*	*Unit*	*Rate*	*Amount*
1	Earthwork in excavation	3540.1	cu m	80 ₹/ cu m	283208 ₹
2	Dry laterite masonry	188.87	cu m	3000 ₹/ cu m	566610 ₹
3	Laterite masonry in cement mortar (1:6)	174.56	cu m	3500 ₹/ cu m	610960 ₹
4	Earthwork in filling	2560.65	cu m	99 ₹/ cu m	253504 ₹
5	Brickwork	7.92	cu m	4393 ₹/ cu m	34792.56 ₹
Total cost					1749074.56 ₹
Contingencies and work charged establishment (5%)					87453.73 ₹
Total cost					**1836528.29 ₹**

4.10 USE OF COMPUTER SOFTWARE AND PROGRAMS IN ESTIMATION

Preparing a detailed estimate is a complex task involving cost estimation for different materials and components. The hand calculations for the thorough estimate are very complex for large projects as it is very difficult to handle and manage a large amount of data. Computer software and programs can help in managing the vast data and calculations in an efficient manner. The major advantages of the computer software and programs for estimating and costing are listed as follows:

(1) The software and computer programs consist of the latest material rate. Computer software and programs can help in the reduction of the cost of construction showing the probable opportunities for cost reduction.

(2) Software and computer programs can handle the vast data and reduce errors in calculations. It eliminates the possibility of human errors. The accuracy of the estimation by software and computer programs is comparatively high.

(3) The use of computer software and programs saves time. The estimate can be prepared rapidly using the available formats and rates on the software.

(4) The estimates prepared by the computer programs and software are easy to analyze and present. The computer programs may provide a graphical interface for better understanding.

(5) The use of software reduces the conflicts among the stakeholders. The data is centralized which enhances the communication among the stakeholders.

(6) The software and computer programs for estimation enhance the clarity in the project and show the professionalism of a firm.

4.11 COMPUTER SOFTWARE AND PROGRAMS FOR ESTIMATION

There are two types of software licenses available in the industry i.e. open access software and subscription-based software. The software that is free to use is known as open-access software while the software that charges some amount of money on an annual or lifetime basis is known as subscription-based software. Some of the famous software for estimation and costing are listed as follows:

(1) **Estimator:** Estimator is a software that is widely used for preparation of the detailed estimates. It helps in preparing accurate bills and detailed estimates. The graphical representations are included in the estimator.

(2) **Contractor foreman:** Contractor foreman is a subscription-based software However, it can be used for on trial basis. It consists of templates for estimates and the latest prices for the materials and labor. It can be integrated with other software like MS Project, Stripe, etc.

(3) **MS Excel templates:** The spreadsheet templates in MS Excel are widely used for estimation and costing. The implementation of the mathematical formulas is easy to implement in Excel spreadsheets. MS Excel is user-friendly and commonly used for preparing detailed estimates.

(4) **Buildsoft:** Buildsoft is an advanced software that is used for the preparation of detailed estimates and helps in project management. The estimates prepared can be verified using the 2D and 3D views of the building in the building information system (BIM).

(5) **Planswift:** It is a cloud-based software. The measurements in the Planswift are very simple and it can be conveniently used for the preparation of paperless estimation. The trial version of this software is free for 14 days.

4.12 CONCLUSIONS FROM THE CHAPTER

The cost of earthwork significantly impacts the cost of the project. The understanding of the earthwork and calculation of earthwork quantities is very important for civil engineering students and professionals. A thorough knowledge of techniques of earthwork estimation may help in reducing the overall cost of the project. The different formulas for earthwork calculation and their practical implementation for highways and canal is discussed in this chapter. A detailed estimate of the septic tank and community well is also provided.

After the completion of this unit, the students will be able to calculate the earthwork quantities for canals and highways. They will solve the problems related to the detailed estimate of the septic tank and well. The error in the detailed estimate of buildings and other civil engineering works may be propagated due to the miscalculation of the rates of the materials and labor. So, an understanding of the rates of material and labor is required for the preparation of accurate estimates. The next chapter deals with the analysis of the rates.

SOLVED NUMERICAL EXAMPLES

Problem 1:

A road is constructed in a city. The chainage varies from 1200 m to 1520 m. Prepare the longitudinal and cross-section for the cutting and filling. The width of the road formation is 9 m. The side slopes provided for cutting are 2:1 and 3:1 for filling. There is a falling gradient of 1 in 200 while the formation is at the ground level for the chainage 1200 m. The reduced levels (RLs) for the different chainages are given in the Table 4.10 below:

Table 4.10: Details of reduced level

S.No.	*Chainage (m)*	*RL of ground (m)*
1	1200	190.20
2	1240	190.55
3	1280	190.80
4	1320	190.10
5	1360	189.90
6	1400	188.95
7	1440	188.75
8	1480	188.10
9	1520	188.15

Prepare a detailed estimate for the earthwork and turfing of side slopes. The excavation, banking and turfing rates are 99 ₹/ cu m, 80 ₹/ cu m and 30 ₹/ sq m respectively.

Solution:

Understand the given data in numerical and plot the longitudinal profile of the section as shown in Figure 4.11 (a).

- The chainage is given from 1200 to 1520 with an interval of 40 m. So, the length of the sections is 40 m each. The RLs are the elevations from mean sea level. The elevation of the ground is given at each chainage. At a chainage of 1200 m, the RL of ground and RL of formation level are the same. However, the RL of the formation line reduces to 1 in 200. This means that if I move 200 m horizontally, the elevation will be reduced by 1 m.
- The formation level at the chainage 1200 m is 190.20 m and it reduces by 1 in 200. The length of one section is 40 m (Chainage 2- Chainage 1= 1240 m -1200 m= 40 m). Change in elevation (RL) for the length of 40 m = $1/200 \times 40 = 0.20\ m$. So, the gradient of formation level is reduced by 0.20 m at each chainage.

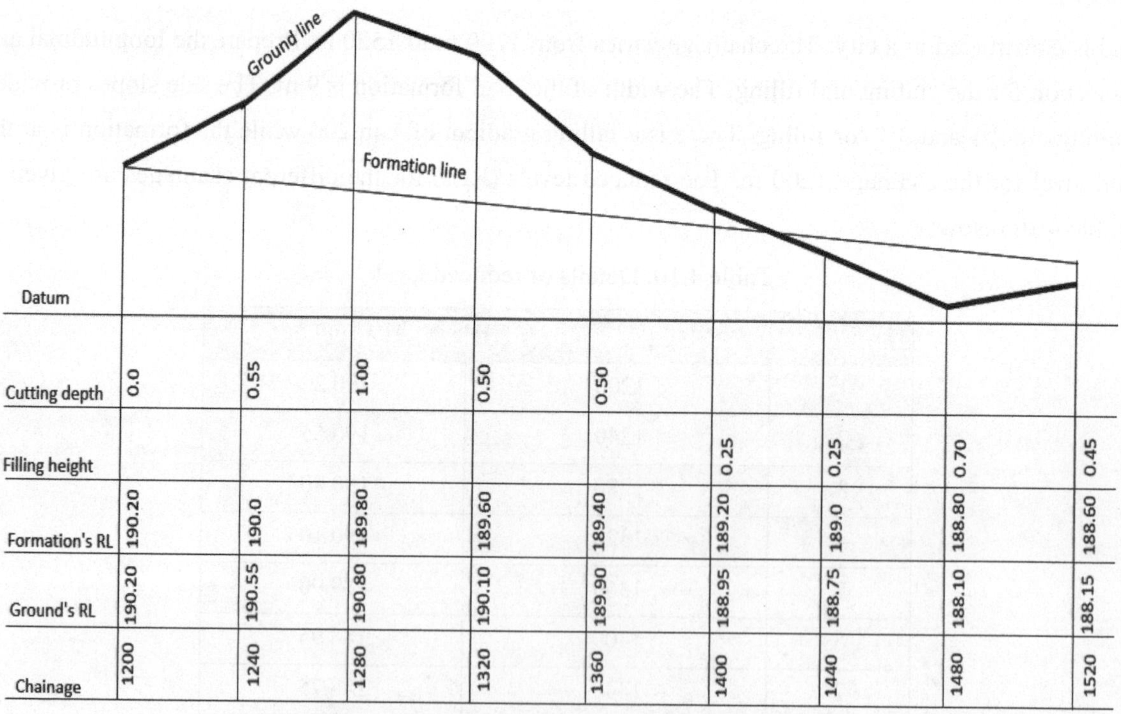

Figure 4.11: (a) Calculations for cutting and filling

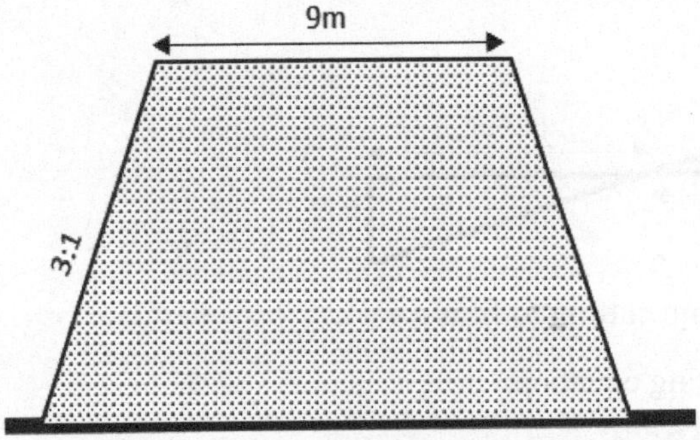

4.11 (b) Section in filling

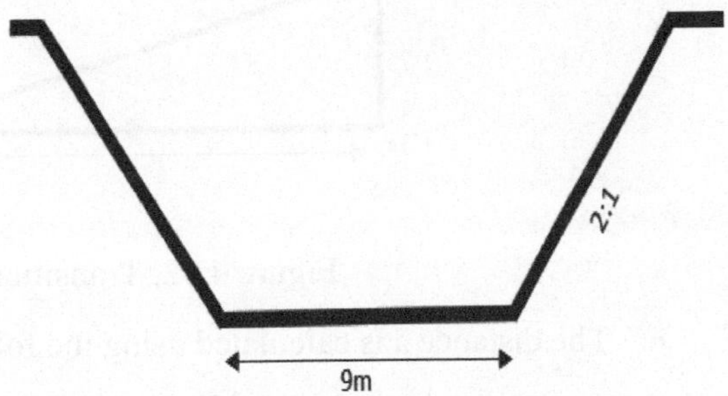

4.11 (c) Section in cutting

- See the example of the calculation for the formation level.
  - Formation level at chainage 1200 m= 190.20
  - Change in elevation= 1/200 × 40= -0.20 (The gradient is reducing)
  - So, the formation level at chainage 1240 m= 190.20- 0.20= 190.0
  - Similarly, for chainage of 1280 m, formation level= 190.0- 0.20= 189.80 m
- The RLs for the ground level are already given in the Table.
- Now, Use the following formula:

  Variation in elevation (E)= RL of formation Level - RL of ground level
  - If E is negative, the cutting is required.
  - If E is positive, the filling or banking is required.
- For calculation of turfing of the slopes keep the following rules in mind:
  - If the cutting is going on, the turfing of the slopes is not required. So, no calculations are needed for the cutting.
  - The breadth of the slope is calculated using the following formula:

    Slope breadth= $h\sqrt{x^2+1}$
    - The area of the sides of slopes is calculated by multiplying the slope breadth by the distance between two sections. There are two slopes, so the area is multiplied by 2.
    - In the given problem, the transition from cutting to filling occurs between the chainage of 1360 m and 1400 m. In this 40 m variation of chainage, the cutting and filling both take place up to certain lengths. This length of cutting and filling needs to be calculated. The figure 4.12 given below shows the transition phase of the earthwork.

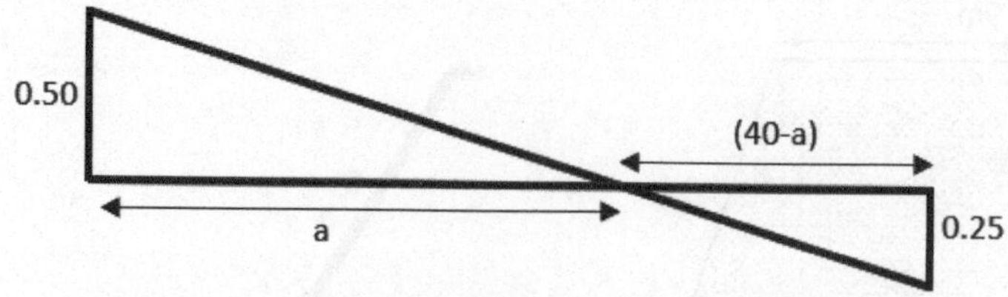

Figure 4.12: Transition from cutting to filling

- The distance a is calculated using the following equation:

$$\frac{0.50}{a} = \frac{0.25}{40 - a}$$
$$0.50\,(40 - a) = 0.25a$$
$$20 - \frac{1}{2}a = \frac{1}{4}a$$
$$\frac{1}{2}a + \frac{1}{4}a = 20$$
$$\frac{2a + 1\,a}{4} = 20$$
$$a = \frac{80}{3} = 26.67 = 27\ m\ (approx)$$

- So, between the chainage from 1360 to 1400, there is cutting work up to 27 m and after that, there is filling for the rest of 23 m.
- For filling, slope is 3:1, so the x is 3.

Table 4.11: Calculation of earthwork

Chainage	Height (H)	Mean Height (H_m)	Central Area (BH_m)	Area of side slopes (xH_m^2)	Total Area ($BH_m + xH_m^2$)	Length between station	Earthwork $(BH_m + xH_m^2)L$	
							Filling (cu m)	Cutting (cu m)
1200	0	-	-	-	-	40	-	-
1240	-0.55	0.275	2.48	0.15	2.63	40	-	105.05
1280	-1	0.775	6.98	1.20	8.18	40	-	327.05
1320	-0.5	0.75	6.75	1.13	7.88	40	-	315.00
1360	-0.5	0.5	4.50	0.50	5.00	40	-	200.00
Passing	0	0.25	2.25	0.13	2.38	40	-	95.00
1400	0.25	0.375	3.38	0.42	3.80	40	151.88	
1440	0.25	0.25	2.25	0.19	2.44	40	97.50	

1480	0.7	0.475	4.28	0.68	4.95	40	198.08	
1520	0.45	0.575	5.18	0.99	6.17	40	246.68	
Total Volume							**694.13 cu m**	**1042.10 cu m**

Table 4.12: Calculations for turfing of slopes

Chainage	*Height (H)*	*Mean Height (H_m)*	*Breadth of slope* $h\sqrt{x^2+1}$	*Length (L)*	*Total Area* $(h\sqrt{x^2+1})\times 2L$
1200	0	During the cutting, the turfing of the slopes is not done. So, the calculations for the cutting are not required.			
1240	-0.55				
1280	-1				
1320	-0.5				
1360	-0.5				
Passing	0				
1400	0.25	0.125	0.395	23	18.17
1440	0.25	0.25	0.790	40	63.2
1480	0.7	0.475	1.502	40	120.16
1520	0.45	0.575	1.818	40	145.44
				Total	**346.97 sq m**

Table 4.13: Abstract of Estimated Cost

Item No.	*Description or particulars of the work*	*Quantity*	*Unit*	*Rate*	*Amount*
1	Earthwork in Excavation	1042.10	cu m	80/ cu m	83368 ₹
2	Earthwork in filling	694.13	cu m	99/ cu m	68719 ₹
3	Turfing of slopes	346.97	sq m	30/ sq m	10410 ₹
Cost of earthwork and turfing					162497 ₹
Contingencies and work charged establishment (5%)					8125 ₹
Total cost					**170622 ₹**

UNIT SUMMARY

- Earthowork may be defined as the removal, placement, redistribution, or compaction of the excavated materials on the construction site.
- Lead may be defined as the horizontal distance from the centre of gravity of the excavation of the earth to the centre of gravity of deposition of the earth.
- Lift may be defined as the vertical distance from the centre of gravity of excavation of the earth to the centre of gravity of deposition of the earth. Figure 4.1 shows the diagram of lead and lift.
- In the mid-sectional area method, the mean height of the section is considered for calculation.
- In the mean section area, the mean of areas of two sections is taken for the calculation of the volume.
- In prismoidal formula the volume is calculated using the following formula:

$$\textit{Volume } (V) = \textit{Distance between section}/3\,[\,(\textit{Area of first section} + \textit{Area of last section}) + 4 \times \textit{Area of odd numbered sections} + 2 \times \textit{Area of even numbered sections}]$$

- In trapezoidal formula the volume is calculated using the following formula:

$$\textit{Volume } (V) = \{\textit{Area of first section} + \textit{Area of last section}/2 + \textit{Area of the rest sections}\} \times \textit{Distance between consecutive sections}$$

- A turfing or pitching of stones is provided along the side-slopes to reduce this soil erosion.
- If the quantity of the cutting earthwork and filling earthwork are nearly equal, it is called 'economical digging'. In this case, the canal is designed with balancing depth.
- If the quantity of earthwork in cutting is more than required for filling, the extra soil is used for spoil banks.
- If the quantity of earthwork in cutting is less than that required for filling, the extra soil is transported from the pits. In this case, no earthwork is considered for spoil banks.
- The septic tank is constructed for the disposal of the night soil in areas with no sewer facilities.
- The night soil is decomposed by an anaerobic reaction in the absence of the air and soaked in the ground by a soak pit.
- The interval of sludge removal may be 1 year to 3 years, however, it is taken two years normally.
- For a small number of users, the capacity of the septic tank is 0.134 cu m per head while for a large number of users, the capacity of the septic tank is 0.07 cu m per head.

- A ventilation pipe must be provided with the septic tank. The minimum diameter of the septic tank is 50 mm for a septic tank.
- The effluent from the septic tank is disposed of in the soak pit or any drain or river. However, the site for the disposal of the effluent should be at least 7.5 m away from any drinking water resource.
- Indian standard code IS: 2470 Part-I provides the details of suitable sizes of the septic tanks to be constructed for different types of buildings.
- The construction of the community wells includes the excavation, brickwork, soil filling, and ring beam work.
- The software and computer programs consist of the latest material rate. Computer software and programs can help in the reduction of the cost of construction showing the probable opportunities for cost reduction.
- Estimator is an open-access software that is widely used for preparation of the detailed estimates. It helps in preparing accurate bills and detailed estimates.
- Buildsoft is an advanced software that is used for the preparation of detailed estimates and helps in project management.

EXERCISE

Multiple Choice Questions

1. If the height at section 1 is 4 m and 6 m at section 2. What will be the height considered in the mid-section area method?

   (a) 10 m (b) 2 m (c) 1.5 m (d) 5 m

2. In which method, the average height of the sections is taken for the calculation of the earthwork?

   (a) Mean-sectional area (b) Mid-sectional area

   (c) Average area method (d) None of the above

3. The area of section 1 is 100 sq m and the area of section 2 is 80 sq m. The distance between these two sections is 20 m. What will be the amount of the earthwork by the mean sectional area method?

   (a) 1200 cu m (b) 1600 cu m (c) 1800 cu m (d) 2400 cu m

   (Hint: $\textbf{\textit{Volume of the earthwork (V)}} = \left(\frac{A_1 + A_2}{2}\right) \times L$**)**

4. The minimum distance of the soak pit from the residential colony is:

   (a) 6 m (b) 6.5 m (c) 7.0 m (d) 7.5 m

5. The minimum diameter for a soak pit is:

   (a) 75 cm (b) 90 cm (c) 95 cm (d) 110 cm

6. The minimum freeboard to be provided for a septic tank is:

   (a) 0.1 m (b) 0.3 m (c) 0.5 m (d) 0.75 m

7. For a small number of users, the capacity of the septic tank is calculated by:

   (a) 0.131 cu m/ head (b) 0.132 cu m/ head (c) 0.133 cu m/ head (d) 0.134 cu m/ head

8. According to the Indian standard code IS: 2470 Part-I, what is the liquid depth of a septic tank for 20 users, which is cleaned at a cleaning interval of 2 years?

   (a) 1.40 m (b) 2.0 m (c) 1.80 m (d) 2.80 m

9. Contractor foreman is an:

   (a) Open access software (b) Unlicensed software
   (c) Subscription-based software (d) None of the above

10. In which software, the estimate can be verified in 2D and 3D plans in BIM:

   (a) Planswift (b) Estimator (c) Buildsoft (d) MS Project

Answers of Multiple-Choice Questions

1. (d), 2. (b), 3. (c), 4. (d), 5. (b), 6. (b), 7. (d), 8. (c), 9. (c), 10. (c)

Short and Long Answer Type Questions

1. Write a short note on the following terms:

   (a) Earthwork in filling (b) Earthwork in cutting
   (c) Lead (d) Lift

2. Explain the procedure of calculating the quantity of the earthwork with the following methods:

   (a) Mean sectional area method (b) Mid-sectional area method
   (c) Prismoidal formula method (d) Trapezoidal formula

3. Write a short note on the area of side slopes.
4. Calculate the quantities of the earthwork for a 200 m-long road. The height at the first section is 2.5 m and the height at the other section is 2.2 m. The formation width is 12 m. The side slope is given 2:1. Calculate the volume using the mid-sectional method, mean section method, and prismoidal method.
5. Write a brief note on the excavation of the canals. Explain the different sections in canal excavation.

6. How do you decide the size of the septic tank? Explain the procedure according to the norms of IS: 2470 Part-I.
7. Explain the procedure of commissioning a new septic tank.
8. Write a short note on the following software:
   (a) Estimator (b) Planswift
   (c) Contractor foreman (d) Buildsoft
9. Calculate the following quantities for the well shown in the figure:
   (a) Earthwork in excavation (b) Earthwork in filling
   (c) Brickwork (d) Laterite masonry work in mortar
   (e) Dry laterite masonry

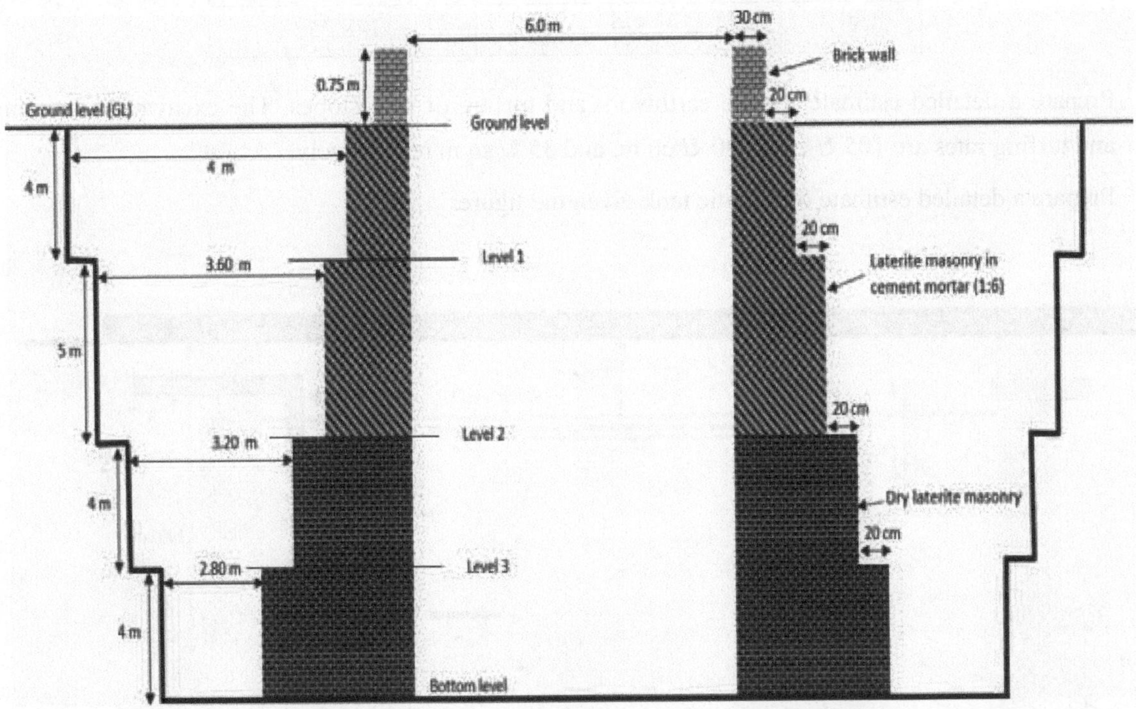

Community well

Also, prepare the abstract of the cost for the given community well.

10. A road is constructed in a rural area. The chainage varies from 200 m to 520 m. Prepare the longitudinal and cross-section for the cutting and filling. The width of the road formation is 12 m. The side slopes provided for cutting are 1.5:1 and 2:1 for filling. There is a falling gradient of 1 in 150 while the formation is at the ground level for the chainage of 200 m. The reduced levels (RLs) for the different chainages are given in the Table below:

Details of reduced level

S.No.	Chainage (m)	RL of ground
1	200	150.35
2	240	150.75
3	280	150.90
4	320	150.30
5	360	149.40
6	400	148.65
7	440	148.75
8	480	148.30
9	520	148.25

Prepare a detailed estimate for the earthwork and turfing of side slopes. The excavation, banking, and turfing rates are 105 ₹/ cu m, 90 ₹/ cu m, and 35 ₹/ sq m respectively.

11. Prepare a detailed estimate of a septic tank given the figure:

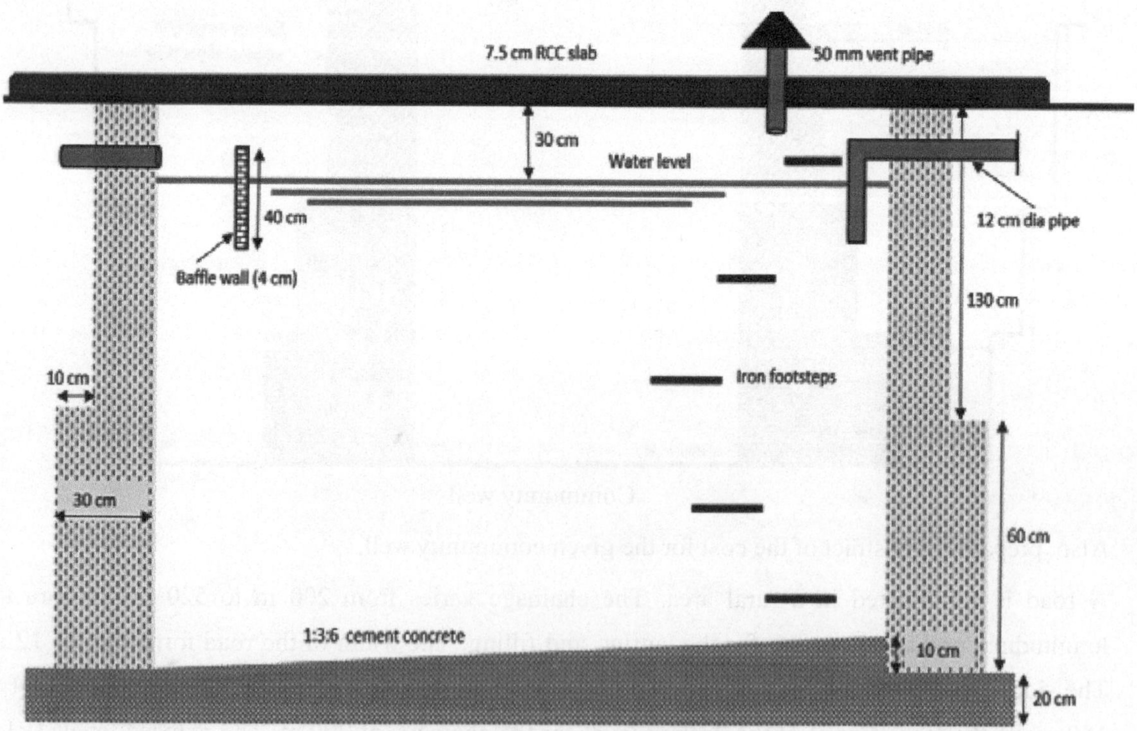

Longitudinal section of septic tank

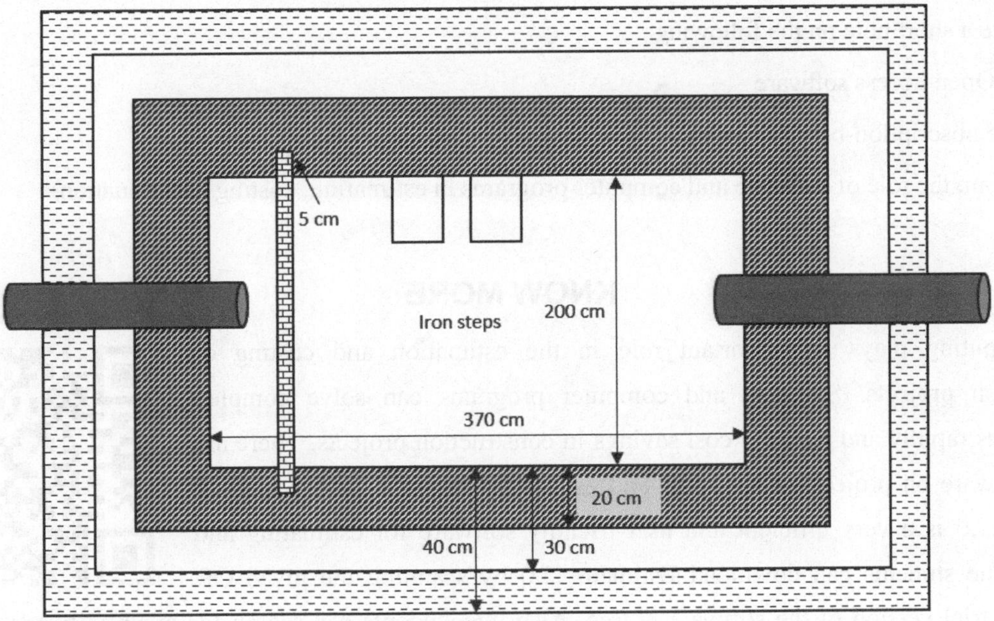

Plan of septic tank

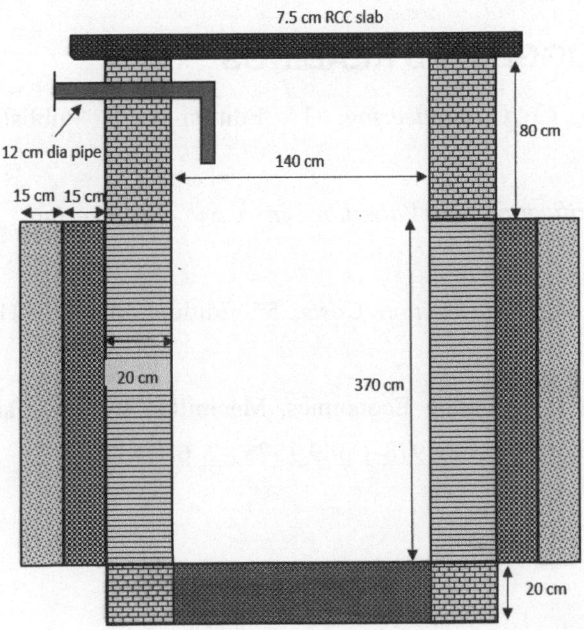

Longitudinal section of soak pit

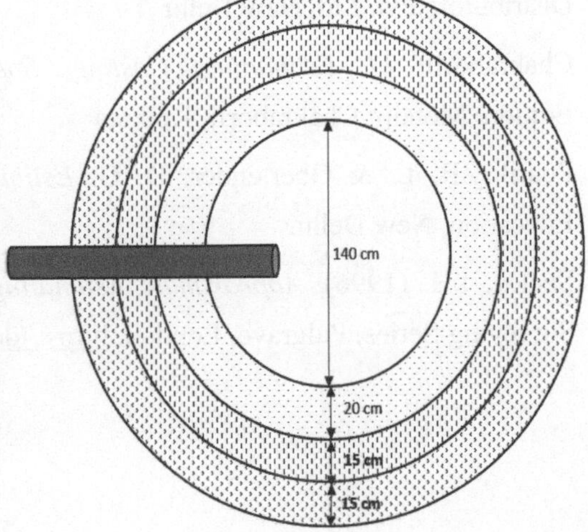

Plan of soak pit

Assume the necessary data. The rates can be taken from the latest schedule of rates of CPWD.

12. Write a short note on the following:
    (a) Open-access software
    (b) Subscription-based software
13. Explain the role of software and computer programs in estimating, costing and valuation.

KNOW MORE

Soft computing plays an important role in the estimation and costing of construction projects. Software and computer programs can solve complex calculations rapidly and result in cost savings in construction projects. There are many software on project management and estimation available in the market. Estimator 2.0 is a very efficient and user-friendly software for estimating and costing. The students can download Estimator 2.0 by scanning the given QR code. The trial version of the software is free. Also, the students can search for other software and use their trial version to learn the estimation and costing on the software.

REFERENCES AND SUGGESTED READINGS

1. Datta, B. N. *Estimating and Costing in Civil Engineering*, 1st Edition UBS Publisher Distributors Pvt. Ltd, New Delhi
2. Chakraborti, *M. Estimating, Costing, Specification & Valuation in Civil Engineering*, 1st Edition, Monojit Chakraborti, Kolkata
3. Peurifoy,R. L. & Oberlender, D. G. *Estimating Construction Costs*, 5th Edition, McGraw Hill Education, New Delhi.
4. Seeley, I.H. (1996). *Approximate Estimating*. In: Building Economics. Macmillan Building and Surveying Series. Palgrave, London. https://doi.org/10.1007/978-1-349-13757-2_6

5

Rate Analysis

UNIT SPECIFICS

Through this unit, we will discuss the following aspects:

- Significance of the rate analysis
- Task work, its significance and task work for different categories of laborers
- The charges of transportation for the materials
- The charges of hire for the machinery and equipment
- Analysis of rates for different items of work

The importance of the rate analysis and its purpose is discussed in detail. The different tables stating the rate of materials, conveyance charges, and hire charges of equipment are discussed to enhance the knowledge about the current rates. The number and types of laborers required for different items of work are also provided. To enhance the understanding of the students and field engineers, some examples of the analysis of rates are elaborated.

This unit consists of exercises with long and short-term questions, multiple-choice questions, a list of references, and suggested reading to make the students learn through practice. Some extra information that may interest the students is provided in the form of QR codes. The students can scan these QR codes to reach the source of further knowledge on the different topics.

RATIONALE

This unit on rate analysis presents the procedure for the analysis of rates and the techniques to decide the final rates of items of work. The categories of the different types of laborers and their daily wages are discussed in detail. The requirements of the labor for different types of work are explained in detail. The transportation charges for materials and hire charges for equipment are presented in tabular format. In the end, the rate analysis for different items of work is performed to enhance the practical knowledge of the students.

The knowledge of analysis of rates is very important for quantity surveyors, estimators, civil engineering students, and professionals. It helps the students understand and decide the realistic rates of the items of work for a construction project. The knowledge of the schedule of rates and the procedure for finalization of rates is essential for civil engineering students and professionals. The knowledge gained through this chapter will help the students in their civil engineering career and they will be able to analyze the rates of items of work accurately which ultimately lead to an accurate detailed estimate.

PRE-REQUISITES

Knowledge of specifications and different work items is required for studying this chapter.

UNIT OUTCOMES

The list of outcomes of this unit is as follows:

U5-O1: Understand the out-turn of the work

U5-O2: Predict the number and types of labor required for different items of work.

U5-O3: Understand different charges to be included in the rates of items

U5-O4: Analyze the rates of items of work

Unit-5 Outcomes	EXPECTED MAPPING WITH COURSE OUTCOMES (1- Weak Correlation; 2- Medium correlation; 3- Strong Correlation)				
	CO-1	CO-2	CO-3	CO-4	CO-5
U5-O1	-	-	-	1	2
U5-O2	-	-	-	1	3
U5-O3	-	-	-	2	3
U5-O4	-	-	-	2	3

5.1 RATE ANALYSIS

The rates of the unit work play a crucial role in estimating and costing of the construction work. If there is an error in the rates of the different items of work, it will lead to unreliable results which will not give the accurate probable cost of construction. So, deciding the rates of the unit work is very important. The study that defines the unit rate of the work by considering the basic requirements like the cost of labor, profit of contractor, etc. is known as the 'rate analysis'.

5.1.1 Purpose of Rate Analysis

The analysis of rates helps in calculating the accurate probable cost of the project by considering the most realistic rates of the material, equipment, labor, etc. The analysis of rates is performed for the following purposes:

(1) The analysis of the rates is done to predict the actual cost of the items. The rates at the local market are also determined for the items.

(2) The rates of the items are checked for viability. The possibilities of bargaining are also considered.

(3) The requirements of the labor and equipment are studied thoroughly and the number of laborers, types of equipment, etc. are also decided.

(4) The rate analysis is performed to determine the use of the equipment, materials, and construction process in an economical way.

(5) Rate analysis is performed to determine the quantity of the extra material or equipment that is not mentioned in the contract but required for the commencement of the work.

(6) The rate analysis helps in revising the schedule of rates based on the changes in material, labor, or technique of construction.

(7) The rate analysis is performed to prepare the contract of labor.

5.1.2 Importance of Rate Analysis

If the rates of the material, labor, or equipment are not accurate, the preparation of an accurate detailed estimate is not possible. So, the rate analysis is very important for the calculation of the realistic probable cost of construction. The importance of the analysis of rates is listed as follows:

(1) The analysis of rates helps in the preparation of an accurate and error-free detailed estimate.

(2) Rate analysis helps the contractor to propose a realistic and economical bid.

(3) The rate analysis helps in comparing the different tenders. Hence, an accurate and realistic analysis enhances the chance of the selection of the proposed tender.

5.2 FIXING UP THE RATES PER UNIT ITEM

The rate per unit item is calculated by adding the different heads. The different heads are discussed as follows:

(1) Quantity and cost of the materials

(2) Rate of earthwork

(3) Cost of labor

(4) Cost of equipment

(5) Water charges

(6) Overhead expenses

(7) Contractor's profit

(1) **Quantity and cost of the materials:** The quantity of the materials is calculated from the drawings and detailed specifications. The specifications help maintain the quality of the materials and the drawings help in the calculation of the quantities of the materials. The rate of the material considered in the estimation is the rate of material delivered at the site. It means that the rate of the material includes the cost at origin, transportation cost, taxes on the cost, etc. The department supplies some of the materials. The profit is not allowed on such types of materials, however, the cost of transportation is provided. Scan QR code to know more.

(2) **Rate of earthwork:** The quantity of the earthwork in filling and cutting is calculated. The cost of lead and lift is also calculated. Lead may be defined as the horizontal distance from the centre of gravity of the excavation of the earth to the centre of gravity of deposition of the earth. The lift may be defined as the vertical distance from the centre of gravity of excavation of the earth to the centre of gravity of deposition of the earth. The extra rates are not provided for the lead up to 30 m and lift up to 1.5 m. In case of lead and lift exceeding the normal values, extra rates are provided. The lift is converted into lead using the following rules:

- If the lift is less than or equal to 3.6 m, the lead is calculated by multiplying the lift by 10.
- If the lift is more than 3.6 m and less than or equal to 6 m, the lead is calculated by squaring the lift and multiplying it by 3.3.
- If the lift is more than 6 m, the lead is calculated by multiplying the lift by 20.

(3) **Cost of labor:** The labor for a construction project is very important for the execution of the physical tasks. The cost of labor includes the wages of the labor, compensations, cost of amenities, etc. The number of laborers required per unit item of work is calculated and multiplied by the cost per laborer to get the cost of labor per unit of work. The working hours for the laborers are 8 hours.

4) **Cost of equipment:** The cost of tools, plants, and equipment is considered while analyzing the rates. The operation charges, maintenance charges, and depreciation charges of the equipment are included in the cost of the equipment. A lump sum amount for the cost of equipment is taken against the unit work.

5) **Water charges:** Water is an essential resource required at every stage of construction. A huge amount of water is required for concrete mixing, curing, washing, cleaning, etc. The municipal corporation charges for the water used in the construction project. The water charges are 1% to 1.5% of the total cost of construction.

(6) **Overhead expenses:** The overhead expenses in a construction project are miscellaneous expenses that should be considered in the analysis of rates. If the overhead charges are not included in the rate analysis, it may become difficult to complete the construction at different stages. The cost of stationary, depreciation of furniture, telephone bills, supervision charges, maintenance and repairing charges of equipment, etc. are included in the overhead expenses. The overhead expenses from 2% to 5% of the total cost are allowed in the construction projects.

(7) **Contractor's profit:** The profit of the contractor is added by the percent of the rate per unit item. The rate of the per unit item is calculated by adding 10% of this cost as the profit of the contractor. If the department supplies the material, the profit on the cost of the material is not allowed. However, 10% of the labor cost is added as the contractor's profit in this case. If it is not mentioned in the contract that the department will supply the material, The contractor's profit is added by 10% of the overall cost of material and labor.

5.3 TASK WORK

The task work or out-turn of the work may be defined as the capacity of a skilled laborer to perform the work in the form of quantity of the work per day. The out-turn of the work depends on the nature, size, height, etc. The out-turn of the work is more in the large cities as compared to the small towns due to the availability of more experienced and skilled laborers. The out-turn of the work by a laborer can be less or more depending on the following factors:

(1) The out-turn of the work depends on the skill of the laborer. If a laborer is more skilled, the out-turn of the work will be more.

(2) The organization and management of the work play an important role in out-turn of the work. If the process of construction and labor on the site are managed efficiently using scientific techniques the out-turn can be increased.

(3) The basic amenities and a good working environment can enhance the performance of the laborers. The out-turn of the work can be increased by providing a good working culture and amenities.

(4) The rewards and incentives for good work can also motivate the laborers to increase their out- turn of work. Table 5.1 shows the materials for the different items of work.

Table 5.1: Materials required for different works

s.no.	description of items	for	quantity	remark
1	Brickwork (Brick size= 20 cm ×10 cm×10 cm)	1 cu m	500 nos	500 nos of bricks are required for 1 cu m construction
2	Dry mortar for brickwork	1 cu m	0.30 cu m	0.30 cu m dry mortar is required for 1 cu m brickwork
3	Stones for rubble stone masonry	1 cu m	1.25 cu m	1.25 cu m stones are required for 1 cu m construction in stone masonry
4	Dry mortar for rubble stone masonry	1 cu m	0.42 cu m	0.42 cu m dry mortar is required for the construction of 1 cu m rubble stone masonry
5	Bricks for brick-ballast	1 cu m	370 nos	370 nos of bricks are required for preparing ballast of 1 cu m.
6	Lime-concrete for brick-ballast	1 cu m	1.05 cu m	-
7	Dry mortar in foundation	1 cu m	0.35 cu m	-
8	Dry mortar in the roof terracing	1 cu m	0.45 cu m	-
9	Bricks in RB work	1 cu m	420 nos	420 nos of bricks are required for construction of 1 cu m reinforced brickwork
10	Cement for cement concrete of mix 1:2:4	1 cu m	0.22cu m	-
11	Sand for cement concrete of mix 1:2:4	1 cu m	0.44 cu m	-
12	Coarse aggregates for cement concrete of mix 1:2:4	1 cu m	0.88 cu m	-
13	Lime for whitewashing (2 coats)	100 sq m	20 Kg	-
14	Ready-mixed paint (1 coat)	100 sq m	5 ltr	-

s.no.	description of items	for	quantity	remark
15	G.C.I sheet for roofing purposes	100 sq m	128 sq m	-
16	Primer (one coat)	100 sq m	6 ltr	-
17	Dry distemper (first coat)	100 sq m	6.5 Kg	-
18	Asbestos cement corrugated sheets	100 sq m	115 sq m	-
19	Timber for the paneled door (4 cm thick shutter)	100 sq m	4.5 cu m	-
20	Timber for fully glazed shutter (thickness= 4 cm)	100 sq m	2.0 cu m	-

5.3.1 Categories of Labor and Their Daily Wages

The laborers are classified based on their skills and expertise. The percentage of the different categories of laborers is also fixed to optimize the construction cost. The labor is classified into the following three categories:

1) Highly skilled labor or skilled first-class labor
2) Skilled labor or skilled second-class labor
3) Unskilled labor

The highly skilled laborers are experts in some work. The highly skilled and skilled laborers are well trained educated laborers who can complete complex physical and mental tasks. They can make independent decisions in different situations during the projects. The skilled first-class laborers are more trained as compared to the skilled second-class laborers.

The unskilled laborers don't have any specialized skills and these types of laborers are assigned simple tasks that may not need any critical thinking or decision-making. 30% of the total skilled labor can be kept as skilled first class labor and the remaining 70% are kept as skilled 2nd class labor. Table 5.2 shows the minimum wages of different types of laborers according to the PWD schedule of Haryana state. However, the rates may vary as per location. The unique codes for each type of labor or job are given by Haryana Public Works Department. These rates are exclusive of GST, contractor's profit and over heads.

Table 5.2: Daily wages of the laborers (As per Haryana PWD SOR 2021)

Unique code	Description	Unit	Rate ([1])
LB001	Bandhani	day	436

Unique code	Description	Unit	Rate ([1])
LB002	Bhisti	day	358
LB003	Blacksmith 1st class	day	376
LB004	Blacksmith 2nd class	day	358
LB005	Carpenter 1st class	day	376
LB006	Carpenter 2nd class	day	358
LB007	Chowkidar	day	376
LB008	Beldar	day	358
LB009	Coolie	day	358
LB010	Fitter (Grade 1)	day	376
LB011	Assistant Fitter or 2nd class Fitter or Fitter (Grade 2)	day	358
LB012	Mali (Semi-Skilled)	day	395
LB013	Glazier	day	376
LB014	Mason (for plaster of Paris work) 1st class	day	458
LB015	Mason 1st class	day	458
LB016	Mason 2nd class	day	415
LB017	Mason (for plain stonework) 2nd class	day	415
LB018	Mason (for ornamental stonework) 1st class	day	458
LB019	Driver for (Road roller, concrete mixer, Trucks, etc.)	day	458
LB020	Mate	day	358
LB021	Sewer man	day	358
LB022	Mistry	day	376
LB023	Painter 1st class	day	376
LB024	Rock Excavator	day	358
LB025	Rock Breaker	day	358
LB026	Rock Hole Driller	day	358
LB027	Stone Chiseler	day	358
LB028	Sprayer (for bitumen, tar etc.)	day	358

Unique code	Description	Unit	Rate (¹)
LB029	Skilled Beldar (for floor rubbing etc.)	day	358
LB030	White Washer	day	358
LB031	Nozzel man/ gun man	day	376
LB032	Mason (average)	day	415
LB033	Carpenter (average)	day	358
LB034	Operator (Pile/ Special machine)	day	358
LB035	Skilled torch operator for laying tack	day	358
LB036	Technician	day	415
LB037	Helper (Technician)	day	358
LB038	Security guard without a gun (8 hours shift duty per day)	day	436
LB039	The security guard with a gun (8 hours shift duty per day)	day	458
LB040	Fitter Beldar	each	358
LB041	Tailor 2nd Class	each	358
LB042	Pump Operators	day	358
LB043	Helper / Chowkidar	day	358
LB044	Electrician	day	250
	IRRIGATION LABOUR		
LB045	Dresser	each	358
LB046	Skilled staff for operating DGPS/Drone for surveying	per day	1000
Horticulture Labor			
LB047	Mali	each	395
LB048	Mali for maintenance	per	10274
LB049	Skilled surveying helper	per day	415
LB050	Total Station Surveyor	per day	766
Electrical Labor			
LB051	Wireman	day	376

Unique code	Description	Unit	Rate ([1])
LB052	Cable jointer	day	376
LB053	Lineman	day	376
LB054	Khallasi	day	358
LB055	Welder	day	376
Job Charges			
JB001	Fabrication of uPVC extruded casement/ sliding windows and doors including drilling holes,	sqm	350
JB002	Installation of uPVC extruded casement/ sliding windows and doors including scaffolding	sqm	350
JB003	Fixing of self-supported arch-shaped galvalume/ Zincalume steel sheet roofing including	sqm	150
JB004	Fixing Hi Rib sheets	sqm	30
JB005	Grinding of granite i/c hiring charges of griding/molding machine.	meter	150
	Horticulture Works		
JB006	Preliminary Watering charges	1000	5
JB007	Ploughing and Dragging swahga with a tractor	per day	5000
JB008	Rolling or ramming of ground turf	sqm	1
PHED Works			
JB009	Repair of Gen. Set if required immediately	per job	500
JB010	Rate of boring 100mm i/d well point	per metre	60

5.3.2 Rules for Calculating the Requirements of the Number of Laborers

The number of laborers required is calculated considering that the maximum work hours are 8 hours. The whole construction work is divided into different categories. The categories of the construction work are further divided into sub-categories based on the different field conditions. The number of laborers of different types are listed for each category and sub-category of the work as discussed below:

(1) **Earthwork (for 10 cu m):** The earthwork is filled and cut. The earthwork is further divided into the following three categories:

- For excavation in ordinary soil with 30 m lead and 1.5 m lift, **3.25 number of mazdoors** are required.
- For excavation in hard soils with 30 m lead and 1.5 m lift, **7.75 number of mazdoors** are required.
- For refilling of excavated soil, **2.33 number of mazdoors** are required.

(2) **Cement concrete work(for 10 cu m):** For the cement concrete work of 10 cu m, the following types of laborers are required:

- Head masons= 0.25 nos
- Masons= 2.5 nos
- Bhisti= 2 nos
- Mazdoor= 10 nos

(3) **Cement concrete work in 25 mm thick Damp Proof Course (DPC) per 100 sq m:** For cement concrete work for a 25 mm thick DPC per 100 sq m, the following types of laborers are required:

- Head masons= 0.5 nos
- Masons= 1 nos
- Bhisti= 1 nos
- Mazdoor= 10 nos

(4) **Laying of lime concrete (20 cu m):** The following types of laborers are required for laying the 20 cu m lime concrete in the foundation:

- Head masons= 0.5 nos
- Masons= 1 nos
- Bhisti= 2 nos
- Mazdoor= 18 nos

(5) **Brickwork (10 cu m):** The labor requirements for the brickwork vary with the component of the structure. The brickwork is further divided into the following categories:

- For first-class brickwork in the foundation and plinth the following types of laborers are required:

- Head masons= 0.5 nos
- Masons= 8 nos
- Bhisti= 2 nos
- Mazdoor= 14 nos

- For first-class brickwork in the superstructure of the groundfloor the following types of laborers are required:
  - Head masons= 0.5 nos
  - Masons= 8 nos
  - Bhisti= 2 nos
  - Mazdoor= 15 nos
- For first-class brickwork in the superstructure of the first floor the following types of laborers are required:
  - Head masons= 0.5 nos
  - Masons= 9 nos
  - Bhisti= 2 nos
  - Mazdoor= 18 nos

(6) **Stonework (20 cu m):** The requirements of the laborer are calculated for 20 cu m of stonework. It is further divided into the following four categories:

- The 20 cu m work of the ashlar masonry in cement mortar or lime mortar requires the following types of laborers:
  - Head masons= 1 no
  - Masons= 50 nos
  - Bhisti= 3 nos
  - Mazdoor= 65 nos
- The 20 cu m work of the random rubble masonry in plinth and foundation in cement mortar or lime mortar requires the following types of laborers:
  - Head masons= 1 no
  - Masons= 20 nos
  - Bhisti= 3 nos
  - Mazdoor= 35 nos

- The 20 cu m work of the random rubble masonry in superstructure in cement mortar requires the following types of laborers:
  - Head masons= 1 no
  - Masons= 22 nos
  - Bhisti= 3 nos
  - Mazdoor= 37 nos
- The 20 cu m work of the coursed rubble masonry in superstructure in cement mortar requires the following types of laborers:
  - Head masons= 1 no
  - Masons= 30 nos
  - Bhisti= 3 nos
  - Mazdoor= 45 nos

(7) **Reinforced brickwork (10 cu m):** The requirements of the laborer are calculated for 10 cu m reinforced brickwork listed as follows:

- Head masons= 0.5 nos
- Masons= 10 nos
- Bhisti= 4 nos
- Mazdoor= 16 nos
- Blacksmith= 4 nos

(8) **Reinforced Cement Concrete (RCC) work:** The requirements of the number of laborers and their types for 10 cu m of RCC work are listed as follows:

- Head masons= 0.5 nos
- Masons= 3 nos
- Bhisti= 4 nos
- Mazdoor= 18 nos

(9) **Plastering (10 sq m):** The requirements of the laborer are calculated for 10 sq m of plastering. The plastering is divided into the two categories listed below:

- The number of laborers required for 12 mm thick plaster for 10 sq m area is given as follows:
  - Head masons= 0.05 no

- ➢ Masons= 1 no
- ➢ Bhisti= 0.1 nos
- ➢ Mazdoor= 1.5 nos

- The pointing work in bricks and cement mortar requires the following types of laborers:
  - ➢ Head masons= 0.05 no
  - ➢ Masons= 1 no
  - ➢ Bhisti= 0.1 nos
  - ➢ Mazdoor= 1 no

(10) **Finishing work (100 sq m):** The finishing work is done in the form of whitewashing, distempering, and painting. The requirements for these three categories are discussed as follows:

- For three coats of whitewashing for 100 sq m area, **1.5 number of painters and 2 number of mazdoors** are required.
- For two coats of distemper and one coat of primer, **6 numbers of painters and 6 numbers of mazdoors** are required.
- For two coats of painting and one coat of primer on a wood or iron surface, **8 number of painters and 8 numbers of mazdoors** are required.

(11) For 100 Kg of reinforcement for RCC work, **1 mazdoor and 1 blacksmith** are required.

(12) For 10 sq m centering and shuttering of the flat surface, **3 carpenters and 4 mazdoors** are required.

(13) For the woodwork of the door frames and window frames, 5.5 mazdoors and 11.1 carpenters are required to complete the 1 cu m work.

(14) For the woodwork of the door shutters and window shutters, 4.5 mazdoors and 16.5 carpenters are required to complete the 10 sq m work.

(15) For CGI roofing, 2 mazdoors, 0.5 blacksmiths, and 1 carpenter are required to complete 10 sq m of work.

(16) **Flooring work (10 sq m):** The requirements of the numbers and types of laborers for the flooring works depend on the type of floors. The flooring works are discussed in the following categories:

- The following types of laborers are required for the construction of a 7.5 cm thick concrete floor with an area of 10 sq m:
  - ➢ Head masons= 0.05 nos
  - ➢ Masons= 1 no
  - ➢ Bhisti= 0.20 nos

- Mazdoor= 2 nos

- The following types of laborers are required for the construction of a 2.5 cm thick cement concrete floor or artificial stone floor with an area of 10 sq m:
    - Head masons= 0.05 nos
    - Masons= 0.8 no
    - Bhisti= 0.20 nos
    - Mazdoor= 1.5 nos
- The following types of laborers are required for the construction of precast terrazzo tiles floor with an area of 10 sq m:
    - Head masons= 0.1 nos
    - Masons= 2 nos
    - Bhisti= 0.1 nos
    - Mazdoor= 2 nos
    - Polisher= 6.5 nos
- The following types of laborers are required for the dado floor with an area of 10 sq m:
    - Head masons= 0.1 nos
    - Masons= 2.5 no
    - Bhisti= 0.5 nos
    - Mazdoor= 4 nos
    - Polisher= 10 nos

5.4 TRANSPORTATION CHARGES OF MATERIALS

The transportation of the materials from one place to another place in the horizontal or vertical direction is required during the construction. Materials like cement, sand, aggregates, etc. must be transported from their source to the site. The following rules are considered for the transportation of the materials:

(1) The transportation and carriage of the materials is performed according to the directions of the Engineer-in-chief. The mode of transportation is provided by the contractor in most cases.

(2) The loading, unloading, stacking, and transporting of the material should be such that there is no loss or damage to the materials or machinery. If the loss or damage of the materials or machinery takes place, the recovery is done at twice of the market rate of the material damaged or lost.

(3) The shortest route is considered for carrying the material. However, if it is not possible to take the shortest route, the recommendations of the Engineer-in-Chief are required for the proposed route.

(4) BIS: 4082 provides recommendations on the stacking and storing of the materials. These recommendations should be followed carefully during the transportation.

(5) The distance between the origin of the material to the construction site at which the material is to be sent is also known as lead. A lead statement provides the cost per unit material including the lead charges. Table 5.3 shows a typical lead statement.

Table 5.3: Lead statement

S.No.	Materials	Unit	Cost at origin (per unit)	Lead in Km	Transportation charges (per Km per unit)	Total transportation charges (per unit)	Total cost (¹/unit)
1	Aggregates	cu m	1500	50	15	750	2250
2	Sand	cu m	580	30	10	300	880
3	Cement	bags	425	local	-	-	-
4	Stones	cu m	1200	50	14	700	1900

(6) The transported material is placed at the least distance from the site. It reduces the further transportation efforts. The transportation charges for the materials are given in Table 5.4. These charges are according to the schedule of rates, Public Works Department (PWD), Haryana. These rates are exclusive of GST, contractor's profit, and overheads.

Table 5.4: Transportation charges for the materials (As per PWD Haryana SOR 2021)

Unique Code	Description	Unit	Rate (¹)
CA001	Good Earth	cum	133
CA002	Dump manure or Sludge	cum	115
CA003	Moorum	cum	106
CA004	Surkhi	cum	106
CA005	Red bajri	cum	106
CA006	Steam coal	tonne	121
CA007	Bricks	1000	283
CA008	Stone aggregate below 40 mm nominal size	cum	106
CA009	Coarse sand	cum	106

Unique Code	Description	Unit	Rate (^{1})
CA010	Timber	cum	121
CA011	Steel	tonne	94
CA012	Stone aggregate 40 mm nominal size and above	cum	115
CA013	Brick tiles	1000	170
CA014	Lime	cum	106
CA015	Cement	tonne	94
CA016	Tar bitumen	tonne	106
CA017	Soling stone & masonry stone	cum	125
CA018	Stone blocks white & red sand stone & kota stone slab	tonne	94
CA019	S.W. pipes100 mm dia	100	142
CA020	S.W. pipes150 mm dia	100	283
CA021	S.W. pipes200 mm dia	100	472
CA022	S.W. pipes250 mm dia	100	810
CA023	S.W. pipes300 mm dia	100	1012
CA024	Brick aggregate	cum	115
CA025	Fly ash	cum	106
CA026	Rubbish	cum	106
CA027	Stone dust	cum	106
CA028	Marble dust and marble chips	cum	106
CA029	G.I. pipes below 100 mm dia	tonne	94
CA030	Stainless Steel pipe below 100 mm dia	tonne	94
CA031	A.C.sheet and accessories	tonne	94
CA032	R.C.C. pipes 100 mm dia	100	232
CA033	R.C.C. pipes 150 mm dia	100	387
CA034	R.C.C. pipes 250 mm dia	100	895
CA035	R.C.C. pipes 300 mm dia	100	1106
CA036	R.C.C. pipes 450 & 500 mm dia	100	2580
CA037	G.I.sheet and accessories	tonne	94
CA038	R.C.C. pipes 600, 700, 750 & 800 mm dia	100	3871

Unique Code	Description	Unit	Rate (^{1})
CA039	R.C.C. pipes 900 mm dia	100	5806
CA040	Plaster of paris	tonne	94
CA041	Cast iron fittings	tonne	94
CA042	Barbed wire	tonne	94
CA043	Spun iron S & S pipes 100 mm dia	100	232
CA044	Spun iron S & S pipes 125 mm dia	100	310
CA045	Spun iron S & S pipes 150 mm dia	100	387
CA046	Spun iron S & S pipes 200 mm dia	100	630
CA047	Spun iron S & S pipes 250 mm dia	100	895
CA048	Spun iron S & S pipes 300 mm dia	100	1106
CA049	Pig lead	tonne	94
CA050	Solvent/ Diesel	quintal	11
CA051	Ductile iron pipes (k7) 100 mm dia	100	232
CA052	Cast iron pipes 150 mm dia	100	387
CA053	Cast iron pipes 200 mm dia	100	630
CA054	Cast iron pipes 250 mm dia	100	895

5.5 HIRE CHARGES OF MACHINERY AND EQUIPMENT

The machinery and equipment used in the construction are either hired or owned by the contractor. The hiring charges for the machinery and equipment are also specified by the state works departments and central works department. Table 5.5 shows the hire charges for machinery and equipment. These rates are exclusive of GST, contractor's profit, and overheads. The unique codes are provided by the Haryana Public Works Department to avoid any confusion.

Table 5.5: Hire charges of machinery and equipment

Unique Code	Description	Unit	Rate (^{1})
PM001	Hire charges of Concrete Mixer 0.25 to 0.40 cum with Hopper	day	800
PM002	Hire charges of Diesel Road Roller - 8 to 10 tonne	day	3000
PM003	Production cost of concrete by batch mix plant	cum	350

Unique Code	Description	Unit	Rate ([1])
PM004	Hire charges of Diesel Truck - 9 tonne	day	2000
PM005	Hire charges of Spraying machine including electric charges	day	250
PM006	Pumping charges of concrete including Hire charges of pump, piping work & accessories	cum	210
PM007	Hire charges of Derrick monkey rope	day	750
PM008	Hire charges of Pump set of capacity 4000 litres/hour	day	700
PM009	Vibrator (Needle type 40 mm)	day	370
PM010	Machine for rubbing of floors	day	300
PM011	Hire and running charges of tipper	day	1700
PM012	Hire and running charges of loader	day	5000
PM013	Hand Grinder for mirror polish	day	250
PM014	Hydraulic Excavator (3D) with driver and fuel	day	7000
PM015	Hire and running charges of light crane	day	3000
PM016	Hire and running charges of bentonite pump	day	3500
PM017	Hire and running charges of crane 20 tonne capacity	day	7000
PM018	Carriage of concrete by transit mixer	km/cum	30
PM019	Generator 250 KVA	day	2500
PM020	Steam curing by using boiler /Heater	cum	500
PM021	Stressing Machine (jack with pump)	day	11500
PM022	Cutting saw machine	day	1350
PM023	Strands Roller machinery for laying strands	day	3500
PM024	Bed master (Pulling strands)	day	3000
PM025	Mobile crane	day	7500
PM026	Tractor with trolley	day	1350
PM027	Air compressor 250 cfm with two leads for pneumatic cutters / hammers	day	1600
PM028	Joint cutting machine with 2-3 blades	day	800
PM029	Cost for crane upto 40 tonne capacity	day	8000
PM030	Water tanker 5000 litre	day	1000

Unique Code	Description	Unit	Rate ([1])
PM031	Air compressor	hour	200
PM032	Cost for crane upto 80 tonne capacity	day	15000
PM033	Concrete Paver finisher with 40 HP Motor and sensor	hour	3000
PM034	Generator 100 KVA/125 KVA (without fuel)	hour	250
PM035	Cost for crane having capacity 50MT	day	8500
PM036	Excavation of Diaphragm wall by Mechanical Grab	sqm	1300
PM037	Hire charges of diesel truck - 9 tonne (witout POL)	day	1500
PM038	Using cost of Ultra Violet Radiation tube	hour	189
PM039	Compressor, gun, rubber pipes & other accessories- hire charge of plant & machinery i/c	day	4000
PM040	Hire Charges of Suction Jeting machine 2200 PSI machine i/c POL and operator	day	36000
PM041	Hire charges of Drill machine upto 30 mm dia	day	160
PM042	Hire charges of sand blasting equipment	day	160
PM043	Hire charges of compressor	day	420
PM044	Welding charges of shear key to existing reinforcement	each	2
PM045	Hire charges of plant and Machinery that can inject 350 kg/day	day	200
PM046	Hire charges for spray pump (Horticulture)	hour	20
PM047	Hire charges brush cutter	hour	100
PM048	Hiring of a tractor with shrub master	hour	450
Hiring instruments for Irrigation works			
PM049	Hire charges of DGPS	per day	2000
PM050	Transport Vehicle i/c fuel	per day	1500
PM051	Lodging Boarding charges for skilled staff	per day	1500
PM052	Hire charges Drone	per day	12000
PM053	Processing of data and printing of drawings	per km	3500
PM054	Hire charges of Total Station and DGS	per day	2000
PM055	Hire charges for JCB with a bucket capacity of 0.4 cum with fuel and driver	per day	6000

Unique Code	Description	Unit	Rate (¹)
PM056	Hire charges for JCB with a bucket capacity of 0.4 to 0.75 cum with fuel and driver	per day	8000
PM057	Hire charges for JCB with bucket capacity more than 0.75 cum with fuel and driver	per day	10000
PM058	Hire charge of the stitching machine	per day	50
PM059	Hire charge of manual trolley	par day	100
PM060	Hire charge of the generator 3 KVA	per hour	40
PM061	Hire charge of 40 quintals. Boat	per day	1400
PM062	Hiring & Running Charges of Electric Cutter	per day	150
PM063	Runing charge of pump (3.5 kilolitre)	per kilolitre	80

5.6 PROCEDURE FOR RATE ANALYSIS

The rate analysis may be defined as the systematic process of determining the rates of items of work. The whole work is divided into the items of the work and their cost is calculated by adding the different types of cost. The detailed procedure for rate analysis is listed as follows:

(1) The work is defined at the beginning of the commencement of the project. The nature of the task, the requirements of the drawings, specifications, laborers, etc. are decided at this stage.

(2) The data required for the rate analysis is collected. The materials are listed for the different items of the work. The specifications and schedule of rate (SOR) are also collected. The equipment and machinery required for the project are also listed.

(3) The cost of the materials is calculated using the schedule of rates and specifications. The transportation charges, overhead charges, taxes, and contractor's profit are also considered as previously discussed.

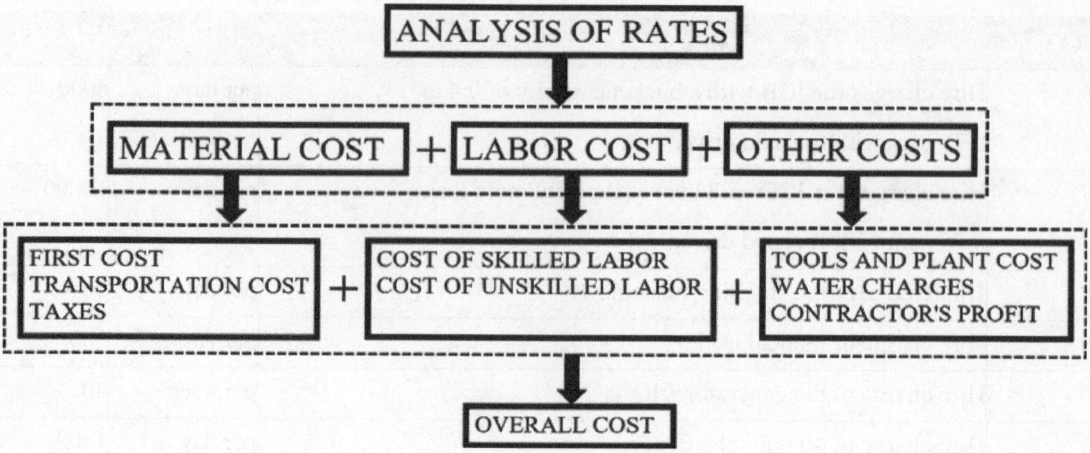

Figure 5.1: Analysis of rates

(4) The out-turn of the work and the number of laborers required at the different stages are determined. The cost of labor is calculated based on the number of laborers and the types required at the different stages of the work.

(5) The charges for hiring the machinery and equipment are calculated using the latest schedule of rates.

(6) The contingencies and overheads are calculated and added to the cost. The total cost is calculated by adding the cost of the equipment, materials, labor, overheads, etc.

(7) The prepared rates are reviewed and adjusted further to finalize the rates. Figure 5.1 shows the flow chart of the analysis of rates.

5.7 PREPARING THE RATE ANALYSIS FOR DIFFERENT WORKS

The rate analysis for the different components of the construction is performed. Table 1 provides the details of the material required. The rate of materials can be obtained from the schedule of rates from the Public Works Department or the competitive authority. It is not possible to provide all the rates of the schedule, however, some rates of materials are provided in **Appendix** at the end of this book. The number and types of laborers required can be obtained from the section 5.3.2.

5.7.1 Rate Analysis for Earthwork

Let's consider that the earthwork at a site is done in foundation. The soil is ordinary. The soil that can be dug with equipment like shovels, spades, etc. is known as ordinary soil. The lift is less than 1.5 m and the lead is less than 30 m. Let's analyze the rates of the earthwork for 10 cu m excavation.

Table 5.6: Rate analysis for earthwork

Description	Quantity (cu m)	Rate	Amount (₹)
Labor		358 ₹/ day	
Mazdoor (Beldar)	3.25		1163.50
Total			1163.50
Contractor's profit (10%)			116.35
Grand total			1279.85
Rate per unit cu m			$\frac{1279.85}{10}$ = **127.9 ₹ per cu m**

mportant note: It is important to note that if the lift and lead are more than normal, we will have to ncrease the number of laborers. For every additional 30 m lead beyond the normal lead, 0.5 mazdoors re added; similarly, for every additional 1.5 m lift, 0.5 mazdoors are added.

5.7.2 Rate Analysis for Concrete Work

Let's consider cement concrete is used in the foundation with a mix of 1:2:4 with sand and stone ballast of 25 mm. Let's analyze the rates for 10 cu m of the cement concrete.

Step 1: Calculate the volume of dry mix:

- Volume of wet concrete= 10 cu m
- Voulme of dry concrete = **50% to 55% more than the wet concrete (It means that to prepare some defined volume of wet concrete, you will be requiring 1.5 to** 1.55 times volume of dry mix.
- Volume of the dry mix= Volume of wet mix+ 50 to 55% of volume of wet mix

  = 1.54×10= 15.4 cu m (We took 54% more volume) Step 2: Calculate the quantities of individual materials:
- The mix is of 1:2:4 (Given)
- No, we will have to calculate the quantities of cement, sand, and stone ballast which are in the ratio of 1:2:4.
- Volume of dry mix= 15.4 cu m
- $Volume\ of\ cement = \frac{\textbf{Total volume}}{\boldsymbol{\Sigma mix\ ratio}} = \frac{15.4}{1+2+4} = \frac{15.4}{7} = 2.2\ cu\ m\ (\ 64\ cement\ bags)$

- Weight of one cement bag= 50 Kg
- Weight of 64 cement bags= 64×50= 3200 Kg= 3.2 tonne

Important note: 1 cu m of cement requires 28.8 cement bags. So, approximately 29 cement bags are required for 1 cu m of cement. For 2.2 cu m cement, 2.2× 29 = 63.8 cement bags are required.

- $Volume\ of\ sand = 2 \times Volume\ of\ cement = 2 \times 2.2 = \mathbf{4.4\ cu\ m}$
- $Volume\ of\ stone\ ballast = 4 \times Volume\ of\ cement = 4 \times 2.2 = \mathbf{8.8\ cu\ m}$

Step 3: Calculate the number of laborers and present the whole data in tabular format.

Now the material rates can be taken from any latest state or central government schedule. The number of laborers and their types can be obtained from section 5.3.2 and Table 2 gives the rate of labor. The analysis of rates for the given problem is presented in Table 5.7:

Table 5.7: Rate analysis for concrete

Description	Quantity	Rate	Amount (₹)
1. Materials			
Cement (OPC 43)	2.2 cu m (3.2 tonne)	4940 per tonne	15800
Sand (Zone III)	4.4 cu m	900 ₹ per cu m	3960
Stone ballast (25 mm)	8.8 cu m	800 ₹ per cu m	7040
2. Labor			
Head mason			
Mason	0.5 nos	458 ₹/ day	229
Mazdoor (Beldar)	1 no	415 ₹/ day	415
Bhisti	18 nos	358 ₹/ day	6444
Miscellaneous	2 nos	358 ₹/ day	716
	Lump sum	1500 ₹	1500
Total			36104.00
Water charges(1 % of the total cost)			361.00
Contractor's profit (10%)			3610.40
Grand total			40075.40 ₹
Rate per unit cu m			$\frac{40075.40}{10} = \mathbf{4007.54}$ ₹ **per cu m**

5.7.3 Rate Analysis for Reinforced Cement Concrete (RCC) Work

Let's consider RCC work with a mix of 1:2:4 in a beam. The shuttering and centering operations are also performed. Let's analyze the rates for 10 cu m of RCC work.

Step 1: Calculate the volume of dry mix:

- Volume of wet concrete= 10 cu m
- Voulme of dry concrete = **50% to 55% more than the wet concrete (It means that to prepare some defined volume of wet concrete, you will be requiring 1.5 to** 1.55 times volume of dry mix.
- Volume of the dry mix= Volume of wet mix+ 50 to 55% of volume of wet mix

  = 1.54×10= 15.4 cu m (We took 54% more volume)

Step 2: Calculate the quantities of individual materials:

- The mix is of 1:2:4 (Given)
- No, we will have to calculate the quantities of cement, sand, and stone ballast which are in the ratio of 1:2:4.
- Volume of dry mix= 15.4 cu m
- $Volume\ of\ cement = \frac{\textbf{\textit{Total volume}}}{\boldsymbol{\Sigma}\textbf{\textit{mix ratio}}} = \frac{\mathbf{15.4}}{\mathbf{1+2+4}} = \frac{\mathbf{15.4}}{\mathbf{7}} = \mathbf{2.2}\ \boldsymbol{cu\ m\ (64\ cement\ bags)}$
- Weight of one cement bag= 50 Kg
- Weight of 64 cement bags= 64×50= 3200 Kg= 3.2 tonne
- $Volume\ of\ sand = 2 \times Volume\ of\ cement = 2 \times 2.2 = \mathbf{4.4}\ \boldsymbol{cu\ m}$
- $Volume\ of\ aggregates = 4 \times Volume\ of\ cement = 4 \times 2.2 = \mathbf{8.8}\ \boldsymbol{cu\ m}$
- Steel bars @ 1% of the volume= 1% of 10 cu m= 0.1 cu m
- Density of steel= 7850 Kg/m^3
- Weight of steel bars in 0.1 cu m= 785 Kg
- Weight of binding wires = 7 Kg (Assumed, there is no particular rule to calculate, however, for 100 Kg steel, 0.9 Kg to 1.3 Kg of binding wires are required.)

Step 3: Calculate the labor:

- See the headings no. 8, 11, and 12 under section 5.3.2. These give details of the type and number of laborers required for RCC work, steel work, centring, and shuttering work. Now, present the data in tabular format.

- **Reinforced Cement Concrete (RCC) work:** The requirements of the number of laborers and their types for 10 cu m of RCC work are listed as follows:
  - Head masons= 0.5 nos
  - Masons= 3 nos
  - Bhisti= 4 nos
  - Mazdoor= 18 nos
- For 100 Kg of reinforcement for RCC work, **1 mazdoor and 1 blacksmith** are required.
- For 10 sq m centering and shuttering of the flat surface, **3 carpenters and 4 mazdoors** are required.

Table 5.8: Rate analysis for RCC work

Description	Quantity	Rate	Amount (1)
1. Materials			
Cement (OPC 43)	2.2 cu m (3.2 tonne)	4940 ₹ per tonne	15800
Sand (Zone III)	4.4 cu m	900 ₹ per cu m 800	3960
Aggregates (20 mm)	8.8 cu m	₹ per cu m 4900 ₹ per quintal	7040
Steel	0.1 cu m (7.85 Quintal)	55 ₹ per Kg	38465
Binding wires	7 Kg		385
			Total= 65650 ₹
2. Labor			
Head mason		458 ₹/ day	229
Mason	0.5 nos	415 ₹/ day	1245
Mazdoor (Beldar)	3 nos	358 ₹/ day	3222
Mazdoor (Coolie)	9 nos	358 ₹/ day	3222
Bhisti	9 nos	358 ₹/ day	1432
Miscellaneous	4 nos Lump sum	1500 ₹	1500
3. Reinforcement			**Total= 10850 ₹**
	8 nos		
Mazdoor	8 nos Lumpsum	358 ₹/ day	2864
Blacksmith		376 ₹/ day	3008
Miscellaneous		1500 ₹	1500
			Total= 7372 ₹

4. Centring and shuttering			
Hire charges (5% of materials cost)	0.05 of 65650	3283.5 ₹	3283.5
Mazdoor	10 nos	358 ₹/ day	3580.0
Carpenter	10 nos	376 ₹/ day	3760.0
Nails (10% of hire charges)	0.10 of 3283.5	328.3 ₹	328.3
Miscellaneous	Lumpsum	1500 ₹	1500.0
			Total= 12451.8
Total (Material cost+labor cost+ reinforcement cost+ shuttering cost)			96323.8 ₹
Water charges(1 % of the total cost)			963.23 ₹
Contractor's profit (10%)			9632.3 ₹
Grand total			**106919.33 ₹**
Rate per unit cu m			$= \frac{106919.33 ₹}{10}$ **= 10691.9 ₹ per cu m**

Important note: The calculations of rate analysis are the same for the slabs, beams, and columns. The procedure of analyzing the rates remains the same for these structural members. The rates for the labor and the materials keep on changing with time. So, the schedule of rates provided by the State works department or central works department should be considered.

5.7.4 Rate Analysis for First-Class Brickwork

Let's consider first-class bricks with a size of 20×10×10 cm are used with the mortar of 1:6 mix. Let's analyse the rates for 10 cu m brickwork. The brickwork is done in the superstructure with first class bricks at the ground floor.

Step 1: Calculation of bricks required for the 10 cu m brickwork:

- Size of bricks= 20×10×10 cm
- Volume of one brick= 0.2×0.1×0.1 cu m
- Total volume= 10 cu m
- Number of bricks required = $\frac{10}{0.2\times0.1\times0.1}$ = 5000 bricks

Step 2: Calculation of the amount of brick mortar:

- Actual size of brick= 19×9×9 cm
- Actual volume of one brick= 0.19×0.09×0.09
- Actual volume of 5000 bricks= 5000× 0.19×0.09×0.09= 7.695 cu m
- Volume of mortar= Total volume- actual volume of bricks

  = 10- 7.695= 2.3 cu m (Approximately)

Add an extra 15% mortar to fill the frog and compensate for the wastage.

- Total volume of the wet mortar= $2.3 + \frac{15}{100} \times 2.3 = 2.65\ cu\ m$

Add 33% extra to the wet mortar to get the volume of dry mortar.

- Volume of dry mortar= $2.65 + \frac{33}{100} \times 2.65$= **3.5 cu m**
- Ratio of cement and sand= 1:6
- Volume of cement= $\frac{Volume\ of\ dry\ mortar}{\sum Ratio\ of\ materials} = \frac{3.5}{7} = 0.5\ cu\ m$= **15 bags (Approximately)**

Important note: 1 cu m of cement requires 28.8 cement bags. For 0.5 cu m cement, 0.5× 28.8 = 14.4. So, approximately 15 cement bags are required.

- Volume of sand= 6× volume of cement= 6×0.5= 3.0 cu m Step 3: Calculation of the labor:
- See section 5.3.2 and for the first class brickwork in the superstructure on the ground floor, the following types and number of laborers are required:
  - Head masons= 0.5 nos
  - Masons= 8 nos
  - Bhisti= 2 nos
  - Mazdoor= 15 nos
- The calculations are presented in the tabular format and the charges for the water and contractor's profit are added to get the final cost.
- The rates of the material and labor can be obtained from the schedule rate issued by the state works department or central works department. These rates in the schedule may be updated from time to time.
- Let's use the rate from the Haryana Public Works Department for 2021-22.

Table 5.9: Rate analysis for brickwork

Description	Quantity	Rate	Amount ([1])
1. Materials			
Bricks	5000 nos	6500 ₹ per thousand	32500
Cement (OPC 43)	0.5 cu m (15 bags)	370 ₹ per bag	5550
Sand (Zone III)	3.0 cu m	800 ₹ per cu m	2400
2. Labor			**Total= 40450 ₹**
Head mason	0.5 nos	458 ₹/ day	229
Mason	8 nos	415 ₹/ day	3320
Mazdoor	14 nos	358 ₹/ day	5012
Bhisti	2 nos	358 ₹/ day	716
Miscellaneous	Lump sum	1500 ₹	
			Total= 10777 ₹
Total (Material cost+labor cost)			51227 ₹
Water charges(1.5 % of the total cost)			768.40 ₹
Contractor's profit (10%)			5122 ₹
Grand total			**57117.40 ₹**
Rate per unit cu m			$= \frac{57117.4 ₹}{10}$ **= 5711.7 □ per cu m**

5.8 CONCLUSIONS FROM THE CHAPTER

In this chapter, the analysis of rates of different materials and labor is discussed. The role of rate analysis and its importance is discussed. The information about the out-turn of the labor is provided in detail. The rate analysis provides a way to achieve an accurate probable cost. It also helps in bargaining the rates. The knowledge of rate analysis is helpful for civil engineers and quantity surveyors to calculate realistic rates for the different items of the work. The accurate rates of items of the work will lead to an efficient and accurate detailed estimate. The analysis of rates majorly depends on the schedule of the rates. So, the latest schedule of rates issued by the competitive authorities is required for the successful analysis of rates. It is also suggested to review the rates after analysis and should be revised to get realistic rates.

UNIT SUMMARY

- The study that defines the unit rate of the work by considering the basic requirements like the cost of labor, profit of contractor, etc. is known as the 'rate analysis'.
- The analysis of rates helps in calculating the accurate probable cost of the project by considering the most realistic rates of the material, equipment, labor, etc.
- The quantity of the materials is calculated from the drawings and detailed specifications. The specifications help maintain the quality of the materials and the drawings help in the calculation of the quantities of the materials.
- Lead may be defined as the horizontal distance from the centre of gravity of the excavation of the earth to the centre of gravity of deposition of the earth.
- The lift may be defined as the vertical distance from the centre of gravity of excavation of the earth to the centre of gravity of deposition of the earth.
- The extra rates are not provided for the lead up to 30 m and lift up to 1.5 m. In case of lead and lift exceeding the normal values, extra rates are provided.
- The cost of labor includes the wages of the labor, compensations, cost of amenities, etc.
- The water charges are 1% to 1.5% of the total cost of construction.
- The overhead expenses from 2% to 5% of the total cost are allowed in the construction projects.
- The rate of the per unit item is calculated by adding 10% of this cost as the profit of the contractor.
- If the department supplies the material, the profit on the cost of the material is not allowed.
- The task work or out-turn of the work may be defined as the capacity of a skilled laborer to perform the work in the form of quantity of the work per day.
- The highly skilled and skilled laborers are well trained educated laborers who can complete complex physical and mental tasks. They can make independent decisions in different situations during the projects.
- For excavation in ordinary soil with 30 m lead and 1.5 m lift, 3.25 number of mazdoors are required.
- For 100 Kg of reinforcement for RCC work, 1 mazdoor and 1 blacksmith are required.
- For 10 sq m centering and shuttering of the flat surface, 3 carpenters and 4 mazdoors are required.

- For the woodwork of the door frames and window frames, 5.5 mazdoors and 11.1 carpenters are required to complete the 1 cu m work.
- The transportation and carriage of the materials is performed according to the directions of the Engineer-in-chief. The mode of transportation is provided by the contractor in most cases.
- The loading, unloading, stacking, and transporting of the material should be such that there is no loss or damage to the materials or machinery.

EXERCISES

Multiple Choice Questions

1. Which of the following is required for rate analysis?

   (a) Specifications (b) Drawings (c) Schedule of rates (d) All of the above

2. Rate analysis includes:

   (a) Cost of material (b) Conveyance cost (c) Hire charges (d) All of the above

3. If a lift of 3.3 m is converted into lead, what will be the value for lead?

   (a) 28 m (b) 33 m (c) 38 m (d) 48 m

4. Water charges are taken usually:

   (a) 1% - 1.5% (b) 1.5% - 3% (c) 3% - 5% (d) 5% -7%

5. Contractor''s profit is added at the rate of:

   (a) 5% (b) 7% (c) 10% (d) 15%

6. If the size of the brick is 20×10×10 cm, how many bricks will be required in 1 cu m?

   (a) 250 (b) 500 (c) 550 (d) 1000

   (Hint: $\textit{Number of bricks required} = \frac{1}{0.2\times0.1\times0.1} = 500\ \textit{bricks}$)

7. The task work per day is calculated for:

   (a) 6 hours (b) 8 hours (c) 12 hours (d) 16 hours

8. How many mazdoors are required for the excavation of ordinary soil in normal lift and lead conditions?

   (a) 3 (b) 3.25 (c) 3.5 (d) 4

9. Which code provides recommendations for stacking and storing the materials?

   (a) IS 4025 (b) IRC 4023 (c) BIS 4082 (d) SP 200

10. If a mix of 1:2:4 is used and the volume of the dry mix is 15.4 cu m. What will be the amount of cement required?

(a) 1.1 cu m (b) 2.2 cu m (c) 3.3 cu m (d) 4.4 cu m

(**Hint:** $Volume\ of\ cement = \frac{Total\ volume}{\sum mix\ ratio} = \frac{15.4}{1+2+4}$)

Answers of Multiple-Choice Questions

1. (d), 2. (d), 3. (b), 4. (a), 5. (c), 6. (b), 7. (b), 8. (b), 9. (c), 10. (b)

Short and Long Answer Type Questions

1. Define rate analysis. What is the significance of the rate analysis in estimating and costing?
2. Explain the procedure of fixing the rates per unit item.
3. Write short notes on the following:
   (a) Water charges
   (b) Overhead charges
   (c) Cost of excavation
4. Write a short note on the out-turn of the work.
5. What is a lead statement? Explain with an example.
6. Explain the procedure of the rate analysis.
7. Write short notes on the following:
   (a) Hire charges of machinery
   (b) Transportation charges
   (c) Daily wages of labor
8. Write short notes on the following:
   (a) Labour required for RCC work
   (b) Labour required for CC work
   (c) Labour required for Earthwork
   (d) Labour required for finishing work
9. What do you understand by a schedule of rates? What is the significance of the schedule of rates?

10. Download any schedule of rates from the web and perform the rate analysis for cement concrete used in the foundation with a mix of 1:4:8 with sand and stone ballast of 20 mm. Analyze the rates for 10 cu m of the cement concrete.

    (Hint: See section 5.7.2)

11. Analyze the rates for 10 cu m RCC work with a mix of 1:1.5:3 in a column. Exclude the rates for shuttering operations.

    (Hint: See section 5.7.3)

12. Perform the rate analysis for first-class brickwork with a brick size of 20×10×10 cm. The mortar mix of 1:4 is used. Analyze the rates for 10 cu m brickwork. The brickwork is done in the superstructure with first-class bricks on the second floor.

    (Hint: See section 5.7.4)

13. Perform the rate analysis for first-class brickwork with a brick size of 20×10×10 cm. The mortar mix of 1:8 is used. Analyze the rates for 10 cu m brickwork. The brickwork is done in the foundation with first-class bricks.

    (Hint: See section 5.7.4)

KNOW MORE

The schedule of rates is an integral part of rate analysis. The latest schedule of rates is required for analyzing the rates and finalizing the rates of different items of the work. There are many schedules of rates available for the reference of civil engineers. In this chapter, we took the rates from the schedule of rates of the Public Works Department, Haryana. This schedule rate is published in the year 2021. If you want a copy of the PWD Schedule of Rates, 2021 for studying it further, Scan the QR code. There are other schedule of rates available for the different state and central government departments. The students can search these schedule rates on the web and download them.

REFERENCES AND SUGGESTED READINGS

1. Datta, B. N. *Estimating and Costing in Civil Engineering*, 1st Edition UBS Publisher Distributors Pvt. Ltd, New Delhi

2. Chakraborti, *M. Estimating, Costing, Specification & Valuation in Civil Engineering,* 1st Edition, Monojit Chakraborti, Kolkata

3. Peurifoy,R. L. & Oberlender, D. G. *Estimating Construction Costs*, 5th Edition, McGraw Hill Education, New Delhi.

4. Seeley, I.H. (1996). *Approximate Estimating*. In: Building Economics. Macmillan Building and Surveying Series. Palgrave, London. https://doi.org/10.1007/978-1-349-13757-2_6

5. Haryana P.W.D (2021). *Schedule of Rates, H.S.R 2021*, Public Works Department, Haryana

APPENDIX

BASIC RATE OF MATERIALS (PWD Haryana 2021)			
Unique Code	**Note: These rates are exclusive of GST, contractor's profit and over heads and carriage etc.**	**Unit**	**Rate (INR)**
	Building Materials		
B0001	Blasting powder	kg	40
B0002	Blasting fuse (fuse wire)	each	15
B0003	Safeda ballies 125 mm diameter	metre	40
B0004	Hollock ballies 125 mm diameter	metre	35
B0005	Fly ash	cum	10
B0006	Kerosene oil	litre	50
B0007	Diesel	litre	57
B0008	Petrol	litre	61
B0009	Mobil oil	litre	250
B0010	Chlorpyriphos 20% E.C. / Lindane 20% E.C.	litre	170
B0011	Brick Aggregate (Single size) : 20 mm nominal size	cum	650
B0012	Brick Aggregate (Single size) : 40 mm nominal size	cum	650
B0013	Brick Aggregate (Single size) : 63 mm nominal size	cum	650
B0014	Over burnt (Jhama) Brick Aggregate: 90 mm to 40 mm size	cum	480
B0015	Stone Aggregate (Single size) : 63 mm nominal size	cum	800
B0016	Stone Aggregate (Single size) : 50 mm nominal size	cum	800
B0017	Stone Aggregate (Single size) : 40 mm nominal size	cum	800
B0018	Stone Aggregate (Single size) : 25 mm nominal size	cum	800
B0019	Stone Aggregate (Single size) : 20 mm nominal size	cum	800
B0020	Stone Aggregate (Single size) : 12.5 mm nominal size	cum	800
B0021	Stone Aggregate (Single size) : 10 mm nominal size	cum	800

B0022	Stone Aggregate (Single size) : 06 mm nominal size	cum	800
B0023	Paving bitumen of grade VG-10 of approved quality	tonne	32000
B0024	Bitumen grade PMB - 40	tonne	32500
B0025	Blown type petroleum bitumen of penetration 85/25 of approved quality	tonne	32500
B0026	Bitumen hot sealing compound: grade A	kg	28
B0027	Bitumen solution primer of approved quality	litre	45
B0028	Curing compound	litre	38
B0029	Portland Cement (OPC-43 Grade)	tonne	4940
B0030	Coal (steam)	quintal	440
B0031	Cement Concrete Jali 50 mm thick	sqm	400
B0032	Cement Concrete Jali 40 mm thick	sqm	350
B0033	Cement Concrete Jali 25 mm thick	sqm	275
B0034	Copper plate	kg	527
B0035	Unslaked lime	quintal	300
B0036	Coarse sand (zone III)	cum	900
B0037	Fine sand (zone IV)	cum	900
B0038	Tangri river sand including cost of royality	cum	900
B0039	Sand zone V (Jamuna)	cum	900
B0040	Average rate of Mild steel round bars for reinforcement	quintal	4900
B0041	Twisted steel/ deformed TMT bars Fe-500D	quintal	5000
B0042	Bolts and nuts up to 300 mm in length	quintal	4800
B0043	Bolts and nuts above 300 mm in length	quintal	5100
B0044	Surkhi	cum	600
B0045	Welding by electric plant	cm	2
B0046	Hard drawn steel wire	quintal	5500
B0047	Mild steel flat strap fitting	quintal	5020

B0048	Plum	cum	600
B0049	50 mm thick interlocking paver blocks (M-30)	sqm	360
B0050	60 mm thick interlocking paver blocks (M-30)	sqm	400
B0051	80 mm thick interlocking paver blocks (M-35)	sqm	500
B0052	100 mm thick interlocking paver blocks (M-35)	sqm	560
B0053	Strips-Aluminium fluted 3.15 mm thick and 150 mm wide	metre	278
B0054	Strips Aluminium fluted 3.15 mm thick and 200 mm wide	metre	370
B0055	1 mm thick Stainless Steel Cover plate grade 304	kg	275
B0056	Coupler 16 mm dia	each	30
B0057	Coupler 20 mm dia	each	39
B0059	Coupler 28 mm dia	each	80
B0060	Coupler 32 mm dia	each	110
B0061	Complete Roof Joint of 100 mm	metre	2800
B0062	Complete Roof Joint of 150 mm	metre	3200
B0063	Complete Roof Joint of 200 mm	metre	4000
B0064	Epoxy adhesive	kg	150
B0065	Floor Joint of 100 mm	metre	3100
B0066	Floor Joint of 150 mm	metre	4000
B0067	Floor Joint of 200 mm	metre	5400
B0068	Wall Joint of 100 mm	metre	2400
B0069	Wall Joint of 150 mm	metre	2800
B0070	Wall Joint of 200 mm	metre	3400
B0071	Bentonite of 35 kg per pile	tonne	2960
B0072	Plasticizer / super plasticizer	kg	36
B0073	Wall form panel 1250x500 mm	each	860
B0074	Tie bolt 12 mm dia 100 mm length	each	38
B0075	Tie bolt 12 mm dia 150 mm length	each	48

B0076	Tie bolt 20 mm dia 150 mm length	each	57
B0077	Tie bolt 20 mm dia 225 mm length	each	67
B0078	Spring coil 12 mm	each	15
B0079	Plastic cone 12 mm dia	each	17
B0080	Corner angle 45x45x5 mm 1.50 m long	each	240
B0081	Corner angle 45x45x5 mm 2.50 m long	each	255
B0082	100 mm channel shoulder 2.5 m long	each	910
B0083	Double clip (bridge clip)	each	76
B0084	Single clip	each	59
B0085	M.S. tube 40 mm dia	metre	215
B0086	Wall form panel 1250x450 mm	each	860
B0087	Column clamp 450x1070 mm	each	965
B0088	Prop 2 m (2-3.5 m)	each	635
B0089	Adjustable span ESO+SI (2.35-3.40)	each	1480
B0090	Adjustable telescopic prop 3 m (2.02-3.75 m)	each	955
B0091	Beam clamp 300-380 mm (450-1070 mm)	each set	355
B0092	Prop 4 m	each	910
B0093	Double coupler	each	46
B0094	Water proof ply 12 mm thick	sqm	517
B0095	Stop end tubes for diaphragmwall 600 mm dia.	sqm	5
B0096	Driving end tubes for diaphragm wall 600 mm dia.	sqm	72
B0097	Seam bolts and nuts 6 mm dia and 25 mm long	10 nos.	10
B0098	Fibre reinforced by organic fibres and/or inorganic synthetic fibres cement corrugated sheet 6	sqm	225
B0099	Fibre reinforced by organic fibres and/or inorganic synthetic fibres cement close fitting	metre	210

B0100	Fibre reinforced by organic fibres and/or inorganic synthetic fibres cement corrugate serrated	metre	210
B0101	Fibre reinforced by organic fibres and/or inorganic synthetic fibres cement plain wing	metre	210
B0102	Fibre (high impact poly propelene reinforced) cement unserrated adjustable ridge for hips	metre	210
B0103	Fibre reinforced by organic fibres and/or inorganic synthetic fibres cement corrugated apron	metre	200
B0104	Fibre reinforced by organic fibres and/or inorganic synthetic fibres cement eaves filler piece	each	175
B0105	Fibre reinforced by organic fibres and/or inorganic synthetic fibres cement north light curves	metre	280
B0106	Fibre reinforced by organic fibres and/or inorganic synthetic fibres cement ventilator curves	each	310
B0107	Fibre reinforced by organic fibres and/or inorganic synthetic fibres cement barge boards	metre	400
B0108	Fibre reinforced by organic fibres and/or inorganic synthetic fibres cement ridge finial	pair	165
B0109	Fibre reinforced by organic fibres and/or inorganic synthetic fibres cement special north light	each	555
B0110	Fibre reinforced by organic fibres and/or inorganic synthetic fibres cement S type louvers	each	260
B0111	Multi purpose fibre reinforced by organic fibres and/or inorganic synthetic fibres cement board	sqm	210
B0112	Multi purpose fibre reinforced by organic fibres and/or inorganic synthetic fibres cement board	sqm	220
B0113	6 mm thick heavy duty fiber cement board	sqm	470
B0114	8mm thick heavy duty fiber cement board	sqm	300
B0115	9 mm thick heavy duty fiber cement board	sqm	625
B0116	12.5 mm thick Gypsum plaster board	sqm	170
B0117	6 mm thick mulitipurpose cement bonded wood particle board conforming to IS : 14276	sqm	195

B0118	8 mm thick mulitipurpose cement bonded wood particle board conforming to IS : 14276	sqm	215
B0119	Factory made light weight composite non asbestos fibre reinforced aerated cement	sqm	634
B0124	Bajri	cum	900
B0125	Bamboo 25 mm dia 2.5 metre long	score	1000
B0126	Bamboo 25 mm dia 3.0 metre long	score	1200
B0127	Bhusa	quintal	500
B0128	Bitumen felt fibre base (vegetable or animal):As per IS 7193 Grade I	sqm	70
B0129	Bitumen felt :Type 3 grade 1	sqm	70
B0130	Coal Tar	litre	28
B0131	White face insulating board: 12 mm thick	sqm	235
B0132	Natural colour insulating board: 12 mm thick	sqm	210
B0133	Flame retardant face insulating board: 12 mm thick	sqm	320
B0134	Flame retardant face insulating, Impregnated fibre board 12 mm thick	sqm	375
B0135	Flat pressed 3 layer particle board (medium density) Grade 1, 12 mm thick	sqm	288
B0136	Extra for veneered particle board with Teak veneering on one side and commercial veneering	sqm	230
B0137	Extra for veneered particle board with Commercial veneering on both sides	sqm	155
B0138	Extra for veneered particle board with Teak veneering on both sides	sqm	500
B0139	Integral crystalline slurry	kg	238
B0140	Integral crystalline admixture	kg	272
B0141	Crystalline mortar	kg	215
B0142	Integral crystalline dry shake	kg	360
B0143	Swellable type water stop tape	metre	365
B0144	Primer for swellable type water stop tape	litre	1540
B0145	Polymer modified adhesive mortar	kg	15

B0146	Brick bats	cum	450
B0147	White Cement	tonne	11200
B0148	Marble dust/ powder	cum	1130
B0149	Mud (dry)	cum	165
B0150	Through and bond stone	100 nos.	5000
B0151	Stone for masonry work	cum	1100
B0152	Stone for pitching 15 cm x 22.5 cm	cum	600
B0153	Stone dust	cum	1000
B0154	Common burnt clay F.P.S. (non modular) bricks class designation 7.5	1000 nos.	6600
B0155	Common burnt clay modular bricks class designation 7.5	1000 nos.	6600
B0156	Common burnt clay F.P.S. (non modular)bricks tile class designation 10	1000 nos.	6600
B0157	Common burnt clay modular bricks class designation 12.5	1000 nos.	6600
B0158	F.P.S. (non modular) clay fly ash bricks class designation 7.5	1000 nos.	5000
B0159	Fly ash bricks conforming to I.S. 12894	1000 nos.	4500
B0160	Extruded burnt flyash clay sewer bricks conforming to I.S 4885	1000 nos.	5400
B0161	Calcium Silicate Bricks machine moulded conforming to I.S. 4139	1000 nos.	5400
B0162	Machine moulded perforated common burnt clay FPS (non modular) bricks of class	1000 nos.	6600
B0163	Machine moulded common burnt clay FPS (non modular) bricks of class designation 12.5	1000 nos.	6600
B0164	Machine moulded common burnt clay modular perforated bricks of class designation 12.5	1000 nos.	6600
B0165	Machine moulded common burnt clay tile bricks of class designation	1000	6600

	12.5	nos.	
B0166	Fire Bricks	100 nos.	3818
B0167	Fire Cement	tonne	1850
B0168	RCC Jali 25mm thick	sqm	400
B0169	Autoclaved aerated cement (AAC) blocks	cum	2500
B0170	Decorative plywood 4 mm	sqm	320
B0171	Copper pins 6 mm dia 7.5 cm long	each	10
B0172	Red sand stone block	10 cudm	79
B0173	Red sand stone slab 30 mm thick (un-dressed)	sqm	200
B0174	Red sand stone slab 40 mm thick (un-dressed)	sqm	200
B0175	Red sand stone slab 45 mm to 50 mm thick (un-dressed)	sqm	225
B0176	Red sand stone gang saw cut 30 mm thick	sqm	400
B0177	White sand stone slab 40 mm thick (un-dressed)	sqm	200
B0178	White sand stone slab 75 mm thick (un-dressed)	sqm	900
B0179	Kota stone slab 20 mm to 25 mm thick (semi-polished)	sqm	280
B0180	Kota stone slab 25mm thick (rough chiseled)	sqm	260
B0181	Cutting marble or sand stone slab up to 50 mm thick by mechanical device	metre	10
B0182	15 mm thick Unistone tiles	sqm	700

REFERENCES FOR FURTHER LEARNING

1. Datta, B. N. *Estimating and Costing in Civil Engineering*, 1st Edition UBS Publisher Distributors Pvt. Ltd, New Delhi
2. Chakraborti, M. *Estimating, Costing, Specification & Valuation in Civil Engineering*, 1st Edition, Monojit Chakraborti, Kolkata
3. Peurifoy, R. L. & Oberlender, D. G. *Estimating Construction Costs*, 5th Edition, McGraw Hill Education, New Delhi
4. Singh, G. & Singh, J. *Estimating Costing and Valuation*, 1st Edition, Standard Publishers Distributers, Delhi
5. IS 1200 (All Parts), *Indian Standard, Method of Measurement of Building and Civil Engineering Works,* Bureau of Indian Standards (BIS), New Delhi
6. Government of India, *Central Public Works Department Specifications*, Vol 1, 2019, published under the authority of the Director General, CPWD, New Delhi, printed and distributed by Jain Book Agency (Central), New Delhi
7. Seeley, I.H. (1996). *Approximate Estimating. In: Building Economics. Macmillan Building and Surveying Series.* Palgrave, London. https://doi.org/10.1007/978-1-349-13757-2_6
8. Oh, C.D., Park, C. & Kim, K.J. *(2013) An approximate cost estimation model based on standard quantities of steel box girder bridge substructure.* KSCE Journal of Civil Engineering, 17, 877–885 (2013). https://doi.org/10.1007/s12205-013-0287-z
9. Haryana P.W.D (2021). *Schedule of Rates, H.S.R 2021*, Public Works Department, Haryana

CO AND PO ATTAINMENT TABLE

Course outcomes (COs) for this course can be mapped with the program outcomes (POs) after the completion of the course and a correlation can be made for the attainment of POs to analyze the gap. After proper analysis of the gap in the attainment of POs necessary measures can be taken to overcome the gaps.

Table for CO and PO attainment

Course Outcomes	**Attainment of Programme Outcomes** *(1- Weak Correlation; 2- Medium correlation; 3- Strong Correlation)*						
	PO-1	**PO-2**	**PO-3**	**PO-4**	**PO-5**	**PO-6**	**PO-7**
CO-1							
CO-2							
CO-3							
CO-4							
CO-5							

The data filled in the above table can be used for gap analysis.

INDEX